groundbakers

60+ PLANT-BASED COMFORT FOOD RECIPES
AND 16 LEADERS CHANGING THE FOOD SYSTEM

BY MACKENZIE AND KATHY FELDMAN

groundbaker

[ground-bay-ker]

Noun: An individual who is an originator, innovator, or pioneer in the transformation toward a just and regenerative food system.

Book cover and interior design by Julie Karen Hodgins

Recipe photography by Hannah Kaminsky
Groundbaker portraits by CE

Cover photos by Erin Scott, Jamel Mosely, Nate Ryan, Ryan Fleisher, and World Central Kitchen

Interior photos by Emma Fishman, Erin Scott, Genisia Green, Hoʻokuaʻāina, Jenni Girtman, Melissa Habegger, Nate Ryan, Paige Green, RYBG, Ryan Fleisher, Ryan Forbes, Sana Javeri Kadri, Soul Fire Farm, Su Evers, Taylor Maruyama, The Sioux Chef, Thuy Tran/Rooted Recipes Project, Tom Hopkins, and World Central Kitchen

Stock photos and illustrations from Shutterstock, iStock, and Freepik

Hardcover ISBN: 979-8-9864733-0-7
Softcover ISBN: 979-8-9864733-1-4

First printing edition 2022

Kulani Publishing

groundbakers.com

For Mom: You will forever be my greatest teacher. Thank you for inspiring me to love food that comes from the Earth, and thank you for saying yes when I asked you to write this book with me. This has been the most valuable experience I could ever have asked for. I cherish every second we spent together in the process. *mackenzie*

acknowledgments

It really does take a village to write a book.

Thank you to the wonderful 16 folks who were willing to share their stories, and more importantly, thank you for your work day in and day out and your dedication to making change.

To the fam: Matt and Mel, we will never forget the day in the café in Oakland when we came up with this idea and wrote down a book outline on napkins. We've come so far. Thank you Mel for all of your contributions to this vision and for being a sounding board every step of the way. Thanks Dad for being a taste-tester for years and for never holding back when you thought something needed more spice.

Thank you Day Hummel for being the best intern we could ask for. We are grateful for all of the people who helped along the way: Rachel Bussey, Asha Culhane-Husain, Taylor Maruyama, Courtney Chun, Cheryl Koehler. Thank you Liz Carlisle and Bryant Terry for teaching us how to write a book proposal.

CE you are so talented, and so patient. Thank you for illuminating the Groundbakers through your beautiful artwork. Rachel Trachten, we couldn't have done this without you; thank you for diving in before you knew what you were signing up for.

Hannah Kaminsky, I don't know how we got so lucky to find you. It's been an incredible 5 years working together. Thank you for making this book what it is today. And to Julie Karen Hodgins, you are brilliant. We knew we were missing something huge, but we didn't know where to look or how to find it. Of course this book couldn't have happened without you, but more importantly, you kept us going when we didn't know where else to turn. Thank you for bringing our vision to life.

With so much love and gratitude,
kathy and mackenzie

contents

*
*Recipes with an asterisk
were provided by our*
groundbakers.

Green Gulch Farm, California

about this book

Groundbakers serves up 60+ plant-based comfort food recipes, ranging from breakfast to dessert. We also feature 16 leaders—whom we refer to as Groundbakers—including farmers, teachers, doctors, activists, and chefs. Each of them shares a personal story, a letter to someone who inspired them along their journey, and a recipe for their favorite healthy version of comfort food. We also include information about pesticides, genetically modified seeds, factory farms, land access, and fisheries. Many cookbooks focus solely on the food, but we wanted to take a broader look at how food reaches our plates. We address a variety of factors, including but not limited to: the people who grow food, the seeds used, soil inputs, the way the animals are treated, and much more.

Many cookbooks focus solely on the food, but we wanted to take a broader look at how food reaches your plate.

When we first set out to create this book, our vision was to develop healthy alternatives to our favorite dishes without sacrificing taste. We wanted to explore and tell the stories of interesting people in the fields of food and farming. We wanted to know who and what led them on this journey, and whom they wish to inspire. We didn't intend to make this a plant-based book, but as we dove deeper, we felt it was important to promote the use of animal-free alternatives for the benefit of humans, animals, and the environment.

As we began to substitute plant-based alternatives for our favorite recipes, we realized that vegan food is anything but boring. The flavors are rich and deep. The dishes are fresh, light, spicy, and delicious! Our recipes are always the hit of the party, even among those who have never tried plant-based food before. People are shocked by the flavor of these dishes and the wide combination of herbs and spices.

WHEN IT COMES TO COOKING

Before you begin to prepare a recipe, we encourage you to first read the entire headnote and recipe. Some dishes require advanced preparation (such as soaking beans and nuts), so it helps to plan ahead whenever possible. Our recipes are gluten-free (except for Einkorn Bread) and vegan, but the recipes submitted by the Groundbakers are just healthy versions of their favorite comfort food, so not all of them are gluten-free and vegan. However, we provide an option to make each dish gluten-free or vegan. Everything is vegetarian.

Mama's Food Pantry (page 4) provides a list of the ingredients that we use most often in our cooking. While the pantry doesn't encompass every ingredient you might use in your cooking, it does provide a good starting point for the ingredients most frequently used in this cookbook. We recommend reading through the pantry section to get an overview of the ingredients, as well as to pick up some tips we've provided about each one.

the whys of a plant-based diet

This book is for everybody. Whether you want to incorporate more plants into your diet or learn about our food system, this book can serve as a resource. It's a step-by-step guide for cooking healthier versions of your favorite foods. While you cook, we will share some facts about the ingredients you're cooking with and how your choices can support positive change for yourself and for the planet.

Becoming informed about our food and where it comes from is a *powerful entry point* toward addressing the systemic problems we care about.

The first step to a deeper understanding of our food system is to face the status quo: our world's food system revolves around factory farming, which is inefficient, cruel, and requires an enormous amount of natural resources to maintain. If all of us reduce our consumption of animal products, we can make a positive impact on the air we breathe, the animals that live among us, the farms and people who grow our food, and our overall health. By reducing animal-based products in our diets, communities would conserve an immense amount of water and drastically decrease pollution. By withdrawing support from extractive corporations, we reduce the threats against the environment and wildlife caused by carbon emissions, deforestation, and industrial agriculture.

According to a 2014 study in the United Kingdom, a diet high in animal products creates 2.5 times more greenhouse gas emissions than a plant-based diet. As a result of factory farms, as well as pesticide use and industrial fisheries, our planet's health is declining. And while individual purchases alone can't create the world we want to live in, becoming informed about our food and where it comes from is a powerful entry point toward addressing the systemic problems we care about. We are here to bring attention to these topics and give you some delicious inspiration for healthy, environmentally- and animal-friendly meals!

Lakeside Organic Gardens, California

Hoʻokuaʻāina, Hawaiʻi

mama's Food pantry

Some of the ingredients we use may sound unfamiliar, but once you stock up and start using them, you will see that a lot of our recipes call for similar ingredients. Once you stock your pantry with the basics, a fun trip to the farmers' market can usually complete any of these recipes. It's also enjoyable and educational to get to know the farmers who grow your food and offer it to you in the freshest way possible. Fresh produce starts to lose its nutrients the minute it's harvested, so being able to enjoy local, farm-fresh fruits and veggies is the healthiest way to go!

BUY ORGANIC

It's important to buy produce not treated with synthetic pesticides and herbicides. Not only is it healthier for you and your family, but also for the farmworkers who toil in the fields to bring you this food. When you get to know local farmers, they will often share their methods and beliefs about growing and harvesting food. Some farmers choose to not be certified organic because of the cost and process involved but instead use sustainable practices that sometimes even go beyond the certified organic rules and regulations.

GRAINS AND FLOURS

Many people have developed a wheat intolerance due to the over-hybridization of wheat in our fields today. We are lucky to have access to many alternative grains that are equally delicious and easier to digest than the wheat grown in the U.S. Here are some suggested grains for your pantry:

Rolled Oats and Oat Flour

Oats are a great source of fiber and are packed with minerals such as manganese, selenium, phosphorus, magnesium, and zinc. We use oats throughout the book in recipes that range from smoothies to baked goods.

Almond Flour

This delicious nutty flour is a great choice for cookies and other baked goods (we feature it in our Chocolate Chip Cookies page 194). Made from ground, blanched almonds, this flour has a slightly sweeter taste than most flours, and it's a great source of protein, as well as manganese, vitamin E, magnesium, and copper.

Einkorn Flour

Although Einkorn flour is a type of wheat flour, it originates from Italy, and has not been hybridized like the wheat we grow in the U.S. It does contain some gluten, but the structure is arranged differently, so it may not have the same affect on those with celiac disease or gluten intolerance. Einkorn flour is also 30% higher in protein than modern wheat. We love its delicious nutty flavor and have created an amazing bread recipe (page 139).

Gluten-Free Flour

Today's wheat just isn't what it used to be. In the 1950s, in order to increase yields, scientists began crossbreeding wheat to make it hardier and shorter. This hybridized wheat contains proteins that some people find hard to digest. Some scientists believe that the gluten and other compounds found in today's modern wheat are responsible for the rise of gluten sensitivity among many people. This is why it's exciting to see how many companies are coming out with different variations of gluten-free baking flours, which are a combination of many flours and powders that can be used to replace wheat baking flour. Two of our favorites are Bob's Red Mill All-Purpose Gluten-Free Baking Flour and Arrowhead Mills Organic All-Purpose Gluten-Free Flour.

Cassava Flour

Cassava flour is made from the root of the yucca plant, which contains a significant amount of calcium and vitamin C. The benefit of this super fluffy, airy flour is its mild, neutral taste. Try our Cassava Flour Tortillas on page 149.

Quinoa

The Incas considered this grain sacred and called it the mother of all grains. Many call it a supergrain since it's rich in protein, contains all essential amino acids, and is a good source of fiber, magnesium, and iron. It has a sweet, subtle taste, is easy to digest, and cooks in about 15 minutes.

LEGUMES

Beans

We refer to many foods as "superfoods" these days, but it's actually beans that may be deserving of this title. Dried beans are easy to store and very inexpensive. They're full of nutrients, including copper, folate, iron, magnesium, potassium, and zinc, as well as protein and fiber.

Chickpeas

Many think of chickpeas—also known as garbanzo beans—as beans, when in fact they are part of of the legume family. This versatile legume has an almost buttery flavor and is a nutritional powerhouse, containing more than 80 nutrients! Chickpeas are also high in fiber and protein, which is why we feature them in a range of recipes, including Dad's Hearty Chili (page 68), Chickpea "Tuna" Salad Sandwich (page 116), and Chickpea Blondies (page 205).

Black Beans

Also classified as a legume, black beans are high in protein and fiber, along with iron, phosphorus, calcium, magnesium, manganese, copper, and zinc. We feature these versatile beans in many of our recipes, from Dad's Hearty Chili (page 68), Southwest Salad (page 96), and Sweet Potato Enchiladas (page 112) to our amazing Black Bean Avocado Brownies (page 192).

Lentils

Lentils come in many different colors and are a common staple around the world. Loaded with protein, fiber, thiamine, niacin, folate, iron, phosphorus, potassium, zinc, copper, manganese, and more, they are easy to use as a side dish or a main course. Try our Lentil Walnut Meatless Meat (page 170) for a tasty surprise!

NUTS AND SEEDS

Nuts and seeds have been part of the human diet since the Stone Age. We incorporate them into most of our recipes because of their health benefits and flavor. You will notice a lot of our recipes call for nut milks, which are easy to make at home. Our delicious cheese substitutes are also made from nuts.

Almonds

Almonds provide a rich source of protein, omega-3 fatty acids, vitamins, minerals, and healthy antioxidants. Of all the nuts, they are among the highest in calcium, which along with their potassium content can help in maintaining strong, healthy bones.

Walnuts

Walnuts are high in omega-3 fatty acids, antioxidants, and phytosterols. We use them in many of our recipes, from our Hemp Granola for breakfast (page 43) to our Lentil Walnut Meatless Meat for dinner (page 170).

Cashews

Cashew nuts are found at the bottom of the fruit known as the cashew apple, native to Brazil, but now grown in many parts of the world including Africa, India, and Vietnam. Although we refer to them as nuts, they are actually seeds. They are a powerhouse of proteins and essential minerals, including copper, calcium, magnesium, iron, phosphorus, potassium, and zinc.

We are big fans of adding superfoods into a recipe without sacrificing flavor. You will notice that the four seeds in the next column are sometimes substituted for eggs or listed as an optional ingredient. We sneak these four seeds into as many recipes as possible to give each meal maximum health benefits.

Flaxseeds

We are constantly trying to sneak fiber into our recipes. That's why seeds play an important role. Flaxseeds have 27.3 grams of fiber per 100 grams, making them one of the top 5% richest foods for fiber content.

Hemp Seeds

Hemp seeds are highly nutritious, containing almost as much protein as soybeans. They are one of the few plant-based foods that are a complete source of protein, meaning that they provide all nine essential amino acids. They are also rich in two essential fatty acids, linoleic acid (omega-6) and alpha-linolenic acid (omega-3). In addition, they are a great source of vitamin E and minerals such as phosphorus, potassium, sodium, magnesium, sulfur, calcium, iron, and zinc.

Chia Seeds

Chia seeds were originally grown in Mexico and highly valued for their medicinal and nutritional properties. The word *chia* means "strength" in the Mayan language, and the seeds were thought to provide energy, strength, and endurance. They are also a good source of omega-3 fatty acids, fiber, antioxidants, iron, and calcium. To top off the benefits, just one ounce of chia seeds provides 10 grams of fiber.

Pumpkin Seeds

These protein-rich seeds are rich in magnesium, healthy fats, zinc, iron, and manganese. We incorporate these seeds in many recipes throughout the book, including our Southwestern Salad (page 96), smoothies (page 24), Hemp Granola (page 43), and Sweet Potato Chickpea BBQ Wraps (page 99).

PLANT-BASED MILKS, YOGURTS, AND CHEESE

When you look around the dairy section of your local market, you will notice new non-dairy alternatives popping up almost weekly. We love to experiment with flavors and textures while reaping the health benefits these products have to offer.

Milks

Our rule for store-bought nut milks is the fewer ingredients, the better. It's fun to experiment with the many varieties that stores offer, or even look online at some of the easy make-it-yourself recipes.

Yogurts

We have not yet tried to make plant-based yogurts, but always go for the cashew, almond, or coconut-based yogurts that are now available in most markets.

Cheese and Butter

We have experimented with some tasty dairy substitutes in this book, such as Cauliflower/ Cashew Sour Cream (page 186), Mozzarella Cheese (page 184), Cashew Parmesan Cheese (page 183), and Almond Tofu Ricotta Cheese (page 182). As for store-bought, plant-based cheese, butter, and sour cream, our favorite brands are Miyoko's, Forager, and Kite Hill.

SWEETENERS

Refined sugar is probably the single worst ingredient in the modern diet. It has been associated with serious diseases and has no nutritional value. We use alternatives to refined sugar throughout this book that are not stripped of nutrients, enabling slower glucose absorption.

Medjool Dates

Dates are nature's perfect sugar substitute, not only because they're deliciously sweet, but also because they're unprocessed and offer extra fiber and potassium.

Maple Syrup

When looking for maple syrup, you will notice that there are four different color classes: Golden, Amber, Dark, and Very Dark. Golden is the lightest, has a delicate taste, and is created at the beginning of the new maple season. Amber is produced in the mid-season and is the most popular all-around flavor. As maple season continues, the syrup darkens in color and Dark is produced, which has a slightly more robust flavor. At the end of the season, Very Dark is produced, which has an intense maple flavor and is perfect for cooking and baking. Always check the label to make sure you are buying pure, 100% organic maple syrup, since many maple-flavored syrups are loaded with refined sugar and artificial ingredients.

Coconut Sugar

Using a high amount of any sugar is not great, but we still prefer coconut sugar to regular sugar because of its nutrients and minerals. Coconut sugar is made from coconut palm sap, which is placed under heat until most of the water has evaporated. During this process, the coconut sugar is able to retain nutrients such as vitamin C, potassium, phosphorus, magnesium, calcium, zinc, iron, vitamin B, and copper. The glycemic index measures 35, whereas white sugar ranks between 60 and 70. In 2014, it was also named by the United Nations Food and Agriculture Organization as the most sustainable sweetener in the world. Coconut trees use minimal amounts of water and fuel compared to cane sugar, and they produce coconut sugar for about 20 years.

Ceylon Cinnamon

We love incorporating cinnamon into our recipes because of the rich, sweet, and comforting flavor it adds to foods and beverages. It also contains a high amount of antioxidants, including polyphenols, phenolic acid, and flavonoids, making it a highly beneficial spice.

BAKING SODA AND BAKING POWDER

Baking soda has one ingredient—sodium bicarbonate. It reacts when it comes into contact with acids, like lemon juice, yogurt, or vinegar. This helps the batter to rise. If there is no acidic ingredient, the dough won't rise.

Baking powder contains sodium bicarbonate, plus a second acid. This helps the batter to rise for a longer period of time, making the batter fluffier. We typically use both baking soda and baking powder together to promote the best rise in flours.

HERBS AND SPICES

Conventional herbs and spices are grown with chemical pesticides and fertilizers. To avoid forming lumps, many manufacturers also use anti-clumping agents like sodium aluminosilicate, sodium ferrocyanide, calcium silicate, and silicon dioxide. It's very disturbing that manufacturers do not need to mention anti-clumping agents in the ingredient list. Organic spices may be a little more expensive, but they do not contain anti-caking agents, harmful pesticides, or chemical fertilizers. Instead, organic spice growers use organic manure and biofertilizers. Here is a list of our favorite herbs and spices:

Salt	*Onion powder*	*Chili powder*
Ceylon cinnamon	*Pepper*	*Cayenne pepper*
Vanilla extract	*Red pepper flakes*	*Smoked paprika*
Curry powder	*Oregano*	*Italian seasoning*
Turmeric powder	*Parsley*	*Cumin*
Garlic powder	*Basil*	

With the unfortunate news that 90% of today's salt contains microplastics, we prefer to use Celtic sea salt from France. Based on the studies done, it seems that this salt has been tested and is clear of microplastics.

SPECIALITY SAUCES

Creating a special sauce for a delicious meal calls for a range of ingredients. The list below contains some of the items that we use over and over again, either to combine with other ingredients or to be used on their own to make the perfect sauce.

Harissa sauce	*Apple cider vinegar*
Coconut liquid aminos	*Vegetable broth (low sodium)*
Tamari sauce (low sodium)	*Dijon mustard*
Tahini	

OTHER DRY INGREDIENTS

As you cook your way through this book, it will make it much easier to have products in your pantry that will be used over and over again. Here is a list of items we keep stored and ready to use in many recipes throughout this book.

Active dry yeast	*Chocolate protein powder*
Nutritional yeast	*Vanilla protein powder*
Cacao powder	*Xanthan gum*
Cacao nibs	

* *Though it looks like a nut butter, tahini is made entirely from ground sesame seeds. It's high in protein, contains minerals such as calcium, iron, and magnesium, and is high in vitamin B.*

OILS

When it comes to oils, there is one that we believe should be avoided at all costs, and that is canola oil. The canola plant does not occur naturally in the wild. The oil is from the domesticated rapeseed-oil plant that has been hybridized and in most cases, genetically engineered to improve its nutritional content and increase the tolerance to herbicides. Today, about 90% of the world's canola crop is genetically engineered to resist Roundup®, an herbicide that has been linked to cancer. The Roundup Ready® canola seed is patented by Monsanto (now Bayer), and farmers can be sued for saving the seed or for having "unauthorized" pollen or seeds from a neighbor's GMO canola crop in their field. Since canola is wind-pollinated, and pollen drift is impossible to stop, it is beyond the control of the organic canola farmers to keep these patented contaminants out of their fields.

Coconut Oil

Coconut oil has a unique combination of fatty acids that may have positive effects on your health. However, it is never a good idea to consume too much of a high-calorie food, so moderation is key. Always look for unrefined organic virgin coconut oil.

Olive Oil

Olive oil has been used in the Mediterranean region for many centuries. To avoid high-heat damage, get extra virgin olive oil that is packaged in an opaque container, which will block the light from deteriorating the oil.

TOOLS AND EQUIPMENT

Below are some of the tools we use the most in the kitchen. Not all of these are necessary, but if you cook a lot, it is definitely worth considering an investment in the following equipment!

Steamer Basket

Steaming is a healthier way to cook veggies than boiling in water, which causes a lot of vitamins and nutrients to be lost.

Pressure Cooker/Instant Pot

This is not a necessity, but investing in an Instant Pot can dramatically change your life! Using one can cut down on time it takes to cook beans, potatoes, rice, and more. If you find yourself devoting more and more time to preparing home-cooked meals, then this is a tool you may want to consider.

Food Processor

Although we use our blender in a lot of recipes, a food processor is sometimes better able to chop and mix foods and to create a meat-like texture.

Small Food Processor

This inexpensive product is great for chopping nuts, seeds, and other small ingredients.

Blender

This piece of equipment is an important investment. Most inexpensive blenders will be useful for a majority of our recipes; however, a Vitamix® or something comparable is extremely powerful and durable, and can transform your creations.

Slow Cooker

We have a few slow cooker recipes in this book, and we recommend it for busy people who want to prepare meals ahead of time. Lots of vitamins and nutrients are lost at high-temperature cooking, whereas slow cookers cook food for a longer time at a much lower temperature and are therefore able to preserve more nutrients.

how-to cooking tips

HOW TO MAKE A FLAX EGG

Sometimes plant-based cooking can be challenging because many recipes require eggs. This is why the flax egg is an amazing plant-based alternative, and once you try it, it opens up the door to so many more recipes.

ingredients

1 tablespoon ground flaxseed

2½ tablespoons water

directions

1. In a small bowl, mix ground flaxseed with water and set aside for 5 minutes.
2. The consistency should resemble that of a scrambled egg. Use as a substitution for 1 egg.

Note: You can set aside in the refrigerator for about 10 minutes after mixing for a better consistency.

HOW TO COOK BEANS

As soon as you grow accustomed to cooking with dried beans, you will see how easy it is. Compared to canned beans, dried beans have less sodium, are less expensive, and they offer superior taste and texture. So, grab a bag of dried beans and let's get started!

1. First, soak 1 cup of dried beans in a large bowl of water (the beans will expand) for at least 8 hours (or overnight). You can refrigerate them or leave them out. After the beans have soaked, rinse them well and place them in a medium/large pot with enough water to cover them by 1 inch. As an option, we like to add a 3x5" piece of kombu, a type of seaweed, to the water while cooking the beans. The amino acids in kombu help with digestion and also release beneficial minerals into the cooking water.

2. Cover pot and bring the water to a boil, then reduce the heat to medium to maintain a gentle simmer.

   Here is a cooking chart for beans:

   Small beans
   Black beans and navy beans
   45–60 minutes

   Medium beans
   Great northern, kidney, pinto, and chickpeas
   60–75 minutes

3. Once cooked, drain and rinse beans, and voilà! They are ready to add to a recipe, store in your refrigerator, or even freeze. Beans will stay fresh in the fridge for up to four days or can be stored in the freezer for months.

HOW TO COOK LENTILS

Lentils come in a variety of colors, sizes, shapes, and consistencies, and they can taste very different. It's important to pick the right lentil for your recipe, so both the taste and texture come out perfectly. Here's a quick rundown on the most popular varieties. Always be sure to rinse your lentils thoroughly before cooking.

Brown Lentils

These are the most common of all the lentils and the ones picked to add a "meaty" texture to your meal. We add 2½–3 cups of water to 1 cup of lentils, cover, and bring to a boil. Keep covered and simmer for about 30–35 minutes until tender.

Green Lentils

These lentils are very similar to brown lentils and are cooked the same way. Add 2½–3 cups of water to 1 cup of lentils, cover, and bring to a boil. Simmer for about 30–35 minutes until tender.

Red and Yellow Lentils

These lentils are sold as "split" lentils, meaning they are processed into smaller bits and have more of a soft, almost mushy texture. To cook this variety, add 3 cups water to 1 cup of lentils, cover, and bring to a boil. Simmer for about 15–20 minutes. Drain out any excess water before serving.

HOW TO SOAK NUTS

Nuts are a vital part of many of our recipes, and soaking them first is very important to achieve all of the benefits they have to offer. Soaked nuts have increased nutritional value and are easier to digest. It is also easier to absorb the nutrients from soaked nuts since they are dense and contain nutritional inhibitors that are released during soaking.

Another important reason to soak these nutritional dynamos is that they become softer and easier to blend. Follow the steps below in order to achieve that smooth, creamy, buttery taste and texture in salad dressings, sauces, cheeses, smoothies, and desserts. For soaking:

1. Place nuts in a bowl with room to expand. Pour enough water into the bowl so they are covered by about 2 inches of water.

2. Place in the refrigerator for 4–6 hours or overnight.

3. Once soaked, drain the water, give your nuts a good rinse, and use them immediately or store in the refrigerator for later.

Quick Soak

If you run out of time and forgot to soak your nuts, don't worry! We use our quick soak method frequently and it always works out just fine.

1. Add 1 cup of nuts to a small pot and add enough water to cover by an inch.

2. Cover pot and bring water to a boil on medium heat. As soon as water starts to boil, remove pot from heat, leave covered, and let nuts soak for 15–20 minutes. Drain nuts and use immediately, or store in the refrigerator or freezer for later use.

HOW TO USE A STEAMER BASKET

Steaming is a cooking method that has been used for thousands of years around the world. One of the earliest examples was in China, where perforated baskets made of bamboo were placed on bowls of boiling water to warm and cook food. Steaming is the best cooking method to ensure the preservation of food's vitamins and minerals. Fast and convenient, these baskets are also reasonably priced (starting at $10). We choose to steam rather than boil vegetables, since a lot of the nutrients are lost when submerging veggies in boiling water.

When steaming vegetables, it is important not to overcook. We recommend following a steaming time chart such as *www.thegardeningcook.com/vegetable-steaming-times* since there are different cooking times for different veggies. Here are some good rules to follow:

1. Chop the veggies in uniform pieces, so that all pieces finish cooking at the same time. The thicker the veggie, the longer it will take to cook.

2. Add about 1 inch of water to the bottom of a pot, cover, and bring to a boil. When water is boiling, place your steamer basket in the pot. The water should be below the basket.

3. Add veggies to the basket and cover pot.
   - Broccoli florets: 4–7 minutes
   - Spinach: 3–5 minutes
   - Green beans: 5–8 minutes
   - Carrot slices: 6–8 minutes
   - Cauliflower florets: 6–8 minutes
   - Potatoes: 15–30 minutes, depending on size

4. Test with a fork to make sure your veggies are done. If still hard to pierce, cook one minute longer.

a family story

*"If there is a book you really want to read but
it hasn't been written yet, then you must write it."*

toni morrison

This is the book that I (Mackenzie) desperately wished for when I set out to learn about our food system: A book that offers healthy, delicious recipes, insight into our broken food system, and models of innovation. Inspired by firsthand experience with brilliant farmers, doctors, professors, and community organizers, I joined forces with my mom to create a book of over 60 plant-based comfort food recipes alongside interviews with 16 influential leaders within our food system: Bryant Terry, Dr. Liz Carlisle, Leah Penniman, José Andrés, Sean Sherman, Michel Nischan, Alice Waters, Dr. Daphne Miller, Anna Lappé, Maricela Vega, Dean Wilhelm, Dr. Malia Smith, Dr. Steve Lawenda, Dr. Gail Myers, Aileen Suzara, and Bob Moore.

Each of the featured leaders (Groundbakers, as we call them) provides a recipe as well as a thank-you letter to someone who inspires them. Through their interviews, along with our included food facts, this book provides a foundational education regarding the most pressing issues our food system now faces—shining a light on animal welfare, workers' rights, soil health, and land access for farmers. This is much more than a cookbook. The goal is to truly break new ground within the cookbook category.

Like many good things in this world, this story starts with my mom. She was one of *those* moms: she bought sprouted grain bread that overpowered the sandwiches, she watered down the juice, and she disguised tofu as meat in hopes that my siblings and I would eat it. Healthy food wasn't just an option in our household—it was the only option.

Forever tempted by the forbidden processed foods, I dreamed about Lunchables and white bread. I'd ask my classmates to trade snacks, although nobody ever wanted

my fruit leathers. Weekends promised long-anticipated sleepovers at my best friend's house, most notable for secret opportunities to over-indulge myself with Lucky Charms and Bagel Bites.

I'll admit, I didn't enjoy our unappetizing pantry staples. As I grew up, I came to understand my mom's methods as I learned more about the negative health effects of pesticides, added sugars, and the antibiotics and growth hormones fed to animals. I realized why she restricted sugary cereals and why she drove 40 minutes to the only grocery store on Oahu that carried organic milk. As I got older, I began to appreciate my mom and the commitment she made to maintaining our well-being—not an easy task in our modern day, industrialized system.

In addition, being Native Hawaiian and growing up in Hawai'i shaped how I view the natural world. I learned that with living on an island comes the kuleana—the responsibility—of protecting limited resources and caring for my home. My Native Hawaiian ancestors once fed their communities by using some of the most historically sustainable agricultural practices ever documented. However, Hawai'i has since become "ground zero" for industrial agriculture, and our communities have seen skyrocketing rates of cancer and birth defects from pesticide usage from agrochemical companies. My mom and I have witnessed the impact that corporate agribusiness has on our islands, and simultaneously, the power of the Hawaiian food sovereignty movement to make positive change. The combination of these lived experiences has shaped much of our understanding of what makes a healthy, sustainable, and equitable food system.

Since the time I could read the nutrition facts on packages, my mom and I have embarked together on a journey to investigate our food system. We became curious about what the "organic" label really meant and what went on at these "organic" farms. Were the farmers really using organic methods? How were the farmworkers treated? Did pesticide drift ever occur from the conventional farms nearby? We had so many unanswered questions. Rather than retreating to armchair internet research, we decided to take a three-week road trip from California to Oregon to see some of these farms for ourselves. We talked with many growers, documenting the information we gathered along the way. The more answers we got, the more questions piled up, and our curiosity continued to grow.

In our current industrialized food system, it was inspiring to witness food grown the right way. In a sense, it restored my hope for the future. At the same time, it reinforced the assertion that our work is far from over. It's not enough to simply grow healthy food. The real problem comes down to accessibility for most Americans.

I learned that with living on an island comes the kuleana— the responsibility— of protecting limited resources and caring for my home.

How do we fix that? My mom and I set out again, this time in search of leaders working to make healthy food more attainable within their communities.

Solutions came in many different ways: by fighting for better policies, educating children, starting nonprofits, and writing books. Whether folks were cooking up new ideas or simply *cooking*, leaders were making healthy food more accessible to their communities. As we heard these stories, we were inspired to experiment with our own tasty alternatives to the foods I used to dream about as a kid. Black beans and avocado? A combination for a killer brownie! (Don't believe me? Check out the recipe on page 192.) My path in college eventually led me to a class called "Edible Education." As Michael Pollan made his way onto the stage for the first lecture, titled The Rise of Industrial Agriculture, I felt as though I was destined to be in that very spot.

I started to tear up within a few minutes. My friends looked at me as if I was crazy, but I knew I was meant to be there. The material didn't just speak to me, it resonated deep within my heart.

I knew right then that studying and advocating for healthy food systems was what I wanted to do for the rest of my life.

What better place than Berkeley, California, home to both the Free Speech Movement and one of the first farm-to-table restaurants, Chez Panisse, to continue my journey? With a mixing bowl of cultures, the highest production of food in the U.S., and a powerful role in the farm-to-table movement, California is the epicenter of radical food justice.

I learned from some of the most brilliant minds in the sustainable food space, such as Saru Jayaraman, Shakirah Simley, Eric Schlosser, Raj Patel, Ricardo Salvador, Sam Kass, Marion Nestle, and Mark Bittman. I would leave

Why should the power of food end with cooking? To me, this is where it all begins.

classes still voracious to learn more. Who are the people who grow our food? How are they treated? What do we put in the soil? I passed up nights of partying with friends so that I could read up on the Farm Bill. I was captivated not by drinking and dancing, but by governmental policies and programs that dictate how farmers care for the land.

Pesticides, workers' rights, racial justice, and animal welfare have everything to do with our food, yet nobody seems to mention them in cookbooks. This lack of information does a disservice to readers and eaters alike. We are missing the full picture. Why should the power of food end with cooking? To me, this is where it all begins.

why food facts?

My Hawaiian upbringing meant being surrounded by nature and a historical sense of responsibility to care for the natural world. It helped me see what was wrong with the industrial poisons used by agribusiness corporations to test their genetically modified seeds on our ancient land, usually near Hawaiian Home Lands. I was inspired by the courage and urgency of everyday people standing up to these corporations to protect people and ʻāina (land), given the painfully high rates of cancer and birth defects in our communities. This led me to start an organization called Re:wild Your Campus. We work with college students and groundskeepers across the country to eliminate toxic, cancer-causing herbicides from school grounds. The vision is to expand beyond campuses and influence EPA legislation in order to protect farmworkers, nearby communities, and consumers from toxic pesticides used in agriculture.

Over the years, I have learned a lot about how human health, food, farming, and environmental justice are interconnected. The sad truth is we are so disconnected from our food that when we sit down to eat every night, we hardly, if ever, think of the people who worked so hard to bring that food to our plates. We cannot have a just and healthy food system if everyone and everything are not treated well, from the workers, to the animals, to the soil, to the water.

mackenzie

We wanted to create a cookbook that also shed light on the issues in our food system. We are not teaching you these food facts to create a dismal picture, but rather to inspire you to take action. We provide a list of ways to get involved on page 230 so you can advocate for a healthy, accessible, and just food system for all.

why organic?

Pesticides affect more than just the soil they are sprayed on. Globally, 44% of farmers experience unintentional acute pesticide poisoning (UAPP), defined by the World Health Organization as workers experiencing one or more symptoms and reporting it within 48 hours of contact with the chemicals. There are an estimated 11,000 deaths globally from UAPP each year, not including deaths from intentional poisoning, as 14 million people have died by suicide using pesticides since the Green Revolution in the 1960s. Let that sink in for a moment: 14 million people. For those who were farmers, one explanation is that during the Green Revolution, chemicals and genetically modified seeds were introduced (see page 56), causing farmers to become dependent on purchasing these agricultural inputs. Compounded with unfavorable policies as well as climate stresses that impacted yields, many farmers were driven into overwhelming debt.

The impacts of the Green Revolution, and the unintentional and intentional deaths from pesticides are a cause for distress. It's also important to note that these figures only account for the acute impacts of pesticides, but do not account for chronic, long-term impacts such as cancer, neurological toxicity, birth defects, and infertility. Many folks who have been exposed to pesticides and suffer from a health issue have not linked it to chemical exposure.

Alongside farmers, children are highly susceptible, as their organs and nervous systems are still developing. Pesticides are often sprayed near playgrounds and lawns, putting children at high risk. In fact, we are all at risk. In the air we breathe, the ground we walk on, the water we drink, and the food we eat, we have all been exposed to synthetic pesticides.

In addition to harming human health, pesticides cause a great deal of harm to insects, which are essential for the functioning of ecosystems. Forty percent of invertebrate pollinator species face extinction, and the world's insect species could go extinct within a century, largely due to widespread use of pesticides, specifically neonicotinoids.

pesticides: a climate change issue

Synthetic pesticides often contain petroleum-based ingredients derived from fossil fuels. Synthetic pesticides contribute to climate change by killing soil microbes and reducing the soil's ability to sequester carbon. Additionally, synthetic fertilizers cause harmful algal blooms and dead zones when they reach lakes and oceans, killing aquatic life due to low oxygen levels.

The good news is that there are alternatives. If soil is maintained in a healthy manner, the soil health increases and has an enormous potential to sequester carbon, reduce water usage, prevent flooding, and protect pollinators. This is possible when soil is managed regeneratively and organically. Whether or not they're certified, farmers who use organic, regenerative methods are helping to fight climate change. Visit a farmers' market in your area, ask questions, support local farmers with good practices, and tell your legislators to enact policies that support organic farmers.

For more information, check out our How to Get Involved section on page 230.

mom's immunity-boosting morning water

prep time: 15 minutes • **freeze time:** Overnight • **makes:** 12–14 large cubes or 16–18 medium cubes

Drinking Morning Water is a ritual I began many years ago. Every morning, I place a frozen cube in my hot water and top it off with the additional ingredients. The combination of vitamins, minerals, and antioxidants provides immune-boosting benefits and gives me the energy I need to start my day. *kathy*

what you'll need

2 large silicone ice cube trays

ingredients

3 lemons, washed, cut in quarters, unpeeled

1 large handful of mint, stems trimmed

1 large piece ginger (*approximately 5–6" in length*), skin peeled (*not necessary to peel all skin off*)

1 large piece turmeric (*approximately 3–4" in length*), skin peeled (*not necessary to peel all skin off*)

2 cups water

additional ingredients

dash of cayenne pepper • dash of black pepper • dash of Himalayan salt
dash of cardamom • 1 teaspoon honey

directions

1. Cut lemons into 4 pieces, leaving skin on. Place lemons into blender, along with mint, ginger, and turmeric.
2. Add 2 cups water and blend all ingredients on high for a full minute.
3. Pour about ¼ cup mixture into each cube. Makes about 12 cubes.
4. Freeze for about 10–12 hours until solid.
5. In the morning, boil 1 mug of water and add a lemon cube to the mug. Add a dash of cayenne pepper, black pepper, salt, cardamom, and honey. Sip and enjoy!

* *While enjoying this drink, **stir occasionally** to avoid settling.*

Lemon peel is high in antioxidants, including D-limonene and vitamin C.

blueberry banana pancakes

prep time: 15 minutes • **cook time:** 12 minutes • **makes:** 6 medium pancakes

When I think of my childhood, I remember happy Sunday meetings with my grandma at the local coffee house, where we always conversed over pancakes. Unfortunately, the brunch usually left me feeling sick and stuffed to the brim. We wanted to create a recipe that sends you back to slow weekend mornings during childhood, but helps you feel good when starting your day. These do the trick and are even better than the pancakes I had growing up! *mackenzie*

ingredients

2 ripe bananas (*one sliced*)

1 flax egg (*page 10*)

1 tablespoon coconut oil, melted

½ cup + 1–2 tablespoons nut milk (*for thinning batter*)

¼ cup plant-based vanilla yogurt

1 teaspoon vanilla extract

½ cup gluten-free flour

We suggest Bob's Red Mill Gluten-Free Flour

½ cup oat flour

1 teaspoon baking soda

½ teaspoon cinnamon

pinch of salt

½ cup blueberries

coconut oil for frying

topping ideas

Vegan Fried Egg (page 177)

directions

1. In a medium bowl, mash 1 banana with a fork. Add flax egg, coconut oil, nut milk, yogurt, and vanilla extract. Set aside.

2. In a separate medium bowl, combine flour mixture, baking soda, cinnamon, and salt. Mix well with a fork.

3. Fold in banana mixture with dry ingredients. Thin to your desire with extra nut milk.

4. Grease large skillet with coconut oil and heat to medium/high. Pour ¼ cup batter into skillet for one pancake, and repeat to cook a few pancakes at a time.

5. Add banana slices and 6–8 blueberries to each pancake. Flip after 2–3 minutes. Cook another 2–3 minutes until both sides are golden brown.

6. Repeat until you finish the batter. Serve immediately.

* **Blueberries** have some of the highest antioxidant levels of all fruits and vegetables.

smoothies

pre-prep: 15 minutes for additional options • **prep time**: 15 minutes
before serving: Place in freezer for ½ hour to thicken • **serves**: 2

*L*iving in Hawai'i, home to fresh fruits and warm days, we've made smoothies a staple in our household for as long as we can remember. Over the years, we have been experimenting with more and more nutrient-dense ingredients to make the most delicious and healthy smoothies possible. We couldn't choose just one favorite, so whether it's a quick on-the-go breakfast or a post-workout snack, we have an option for every craving.

Chocolate Mint Smoothie

Cotton Candy Smoothie

optional ingredients

1 tablespoon ground flaxseed

1 tablespoon chia seeds

1 tablespoon hemp seeds

1 tablespoon rolled oats

1 tablespoon pumpkin seeds

¼ cup water

topping ideas

Hemp Granola (*page 43*)

coconut whipped cream

We recommend the So Delicious CocoWhip.

Pumpkin Pie
Smoothie

Blueberry
Cacao
Smoothie

Peanut Butter
and Jelly
Green Smoothie

chocolate mint smoothie

ingredients

2 dates, pitted
2 cups spinach, slightly steamed (*page 12*)
¼ cup mint leaves
1 cup nut milk of choice
¼ avocado
1 tablespoon cacao powder
1 scoop chocolate or green protein powder
1 banana, fresh or frozen
1 teaspoon vanilla extract
pinch of sea salt

for pulsing after blending

1 tablespoon cacao nibs
1 tablespoon chocolate chips

1. If adding additional ingredients, mix them in ¼ cup water (feel free to add in the dates as well, to soften them before blending). Let soak for 10 minutes.
2. Place all ingredients in the blender and blend until thoroughly mixed.
3. Add cacao nibs and chocolate chips to blender and pulse lightly.
4. Pour into 2 cups and place in freezer for 30 minutes to thicken. Feel free to add more nibs and chocolate chips for topping if you have extra.

cotton candy smoothie

ingredients

1 packet frozen pitaya
3–4 dates, pitted
½ cup strawberries, fresh or frozen
1 teaspoon vanilla extract
1 banana, fresh or frozen
1 cup coconut or almond milk

1. If adding additional ingredients, mix them in ¼ cup water (feel free to add the dates in there as well to soften them before blending). Let soak for 10 minutes.
2. Place all ingredients in the blender and blend until thoroughly mixed.
3. Pour into 2 cups and place in freezer for 30 minutes to thicken.

* *Native to Mexico and Central America, **pitaya**, also known as dragon fruit, is becoming more and more popular and is now grown all over the world. This beautiful fruit is also very nutrient-dense, high in fiber and magnesium, and loaded with antioxidants.*

blueberry cacao smoothie

ingredients

1½ cups frozen blueberries
2 dates, pitted
¾ cup nut milk
1 banana, fresh or frozen
1 tablespoon nut butter
1 tablespoon cacao powder
1 tablespoon cacao nibs
1 scoop chocolate protein powder
¼ cup plant-based plain yogurt

for pulsing after blending

1 tablespoon cacao nibs
1 tablespoon chocolate chips

1. If adding additional ingredients, mix them in ¼ cup water (feel free to add the dates in there as well to soften them before blending). Let soak for 10 minutes.
2. Place all ingredients in the blender and blend until thoroughly mixed.
3. Add the cacao nibs and chocolate chips. Pulse 4–5 times to blend lightly.
4. Pour into 2 cups and place in freezer for 30 minutes to thicken.

peanut butter and jelly green smoothie

ingredients

3 dates, pitted
1 cup almond milk
2 bananas, fresh or frozen
1 tablespoon peanut butter (or any nut butter)
1 tablespoon peanut butter powder
2 tablespoons vanilla protein powder
1 cup spinach, slightly steamed (page 12)
6–8 strawberries, fresh or frozen

topping ideas

strawberries, sliced
additional nut butter

1. If adding additional ingredients, mix them in ¼ cup water (feel free to add the dates in there as well to soften them before blending). Let soak for 10 minutes.
2. Place all ingredients in the blender and blend until smooth and creamy.
3. Pour into 2 cups and place in freezer for 30 minutes to thicken. Feel free to top with sliced strawberries and drizzle with nut butter when serving.

* Spinach *is a great source of vitamins and minerals, including vitamins A, C, K1, folic acid, iron, and calcium. We steam our spinach before adding it because it contains oxalic acid, which blocks the absorption of iron and calcium but breaks down under high temperatures.*

pumpkin pie smoothie

pre-prep: Soak cashews or walnuts for 4 hours or overnight, or quick soak (page 12)

ingredients

3 dates
1 tablespoon cashews or walnuts, soaked and drained
1 cup kabocha squash, cooked
½ cup coconut milk
½ cup almond milk
2 bananas, fresh or frozen
1 teaspoon vanilla extract
½ teaspoon ground cinnamon
½ teaspoon pumpkin pie spice
¼ teaspoon turmeric powder

1. If adding additional ingredients, mix them in ¼ cup water (feel free to add the dates in there as well to soften them before blending). Let soak for 10 minutes.
2. Place all ingredients in the blender and blend until thoroughly mixed.
3. Pour into 2 cups and place in freezer for 30 minutes to thicken.

* Kabocha squash, *also known as winter squash, is a rich source of vitamin C and beta carotene, which the body converts into vitamin A.*

dreamy chocolate breakfast shake

pre-prep: 15 minutes for additional options – optional to freeze bananas overnight • *prep time*: 10 minutes

before serving: Place shake in freezer for 30 minutes • *serves*: 2

My grandma always dreamed about a day when a chocolate milkshake could actually be good for you. So, we knew this had to be one of the recipes in the book. After trying about 50 different combinations over the years, we came up with the best shake imaginable—full of healthy fats, vitamins, and protein—that tastes even better than your favorite burger-joint shake! And wait, it gets better … we even recommend having this for breakfast as a great way to start your day! This one's for you, Grandma! *mackenzie*

ingredients

2 medjool dates
1 cup plant-based milk
¼ cup coconut milk
3 bananas, fresh or frozen
1 scoop chocolate protein powder
2 tablespoons cacao powder
1 teaspoon vanilla extract

additional options

1 tablespoon ground flaxseed
1 tablespoon hemp seeds
1 tablespoon chia seeds
1 tablespoon rolled oats
1 tablespoon pumpkin seeds
¼ cup water

topping ideas

coconut whipped cream
We recommend the So Delicious CocoWhip.

cacao nibs
chocolate and/or coconut shavings
Chocolate Sauce (*page 214*)
Chocolate Hazelnut Spread (*page 209*)

directions

1. If adding additional options, mix them in ¼ cup water (feel free to add the dates in there as well to soften them before blending). Let soak for 10 minutes.
2. Add all ingredients into a blender, including any additional options you choose, and blend well.
3. We recommend placing the shake in the freezer for at least 30 minutes before serving.

anna
LAPPÉ

Anna Lappé is a national best-selling author, widely respected advocate for food justice and sustainability, TEDx speaker, and advisor to funders investing in food system transformation.

A recipient of the 2016 James Beard Leadership Award, Anna is the author of *Diet for a Hot Planet*; co-author of *Hope's Edge*, with her mother, Frances Moore Lappé; and co-author of *Grub: Ideas for an Urban Organic Kitchen*, with Bryant Terry. A founding principal of the Small Planet Institute and Small Planet Fund, Lappé has been named one of *Time*'s "Eco" Who's Who, and has been featured in national publications such as the *New York Times*. She is also the founder and strategic advisor of Real Food Media, which uses videos, radio, podcasts, social media, public speaking, writing, campaign strategy, and storytelling for food justice and food sovereignty. Anna also launched and runs the Food Sovereignty Fund of the Panta Rhea Foundation, which among other work helps support adoption of the Good Food Purchasing Program across the United States. In her own words, "I help people understand the incredible power and potential of sustainable food systems to nourish all of us and expose how some of the biggest corporations in the food system try to undermine that understanding."

Anna Lappé grew up having conversations about health, food justice, and sustainability with her mother, Frances Moore Lappé. Frances is a researcher, author, and co-author of 20 books on the topics of world hunger, living democracy, and the environment. While attending Brown University, Anna was inspired by one of her professors, Ted Sizer, a leader in progressive education reform.

Anna got involved with his work as an undergraduate student and focused on progressive public education. She valued public education and wanted it to be protected from the forces for privatization that were starting to gain power at the time. After graduating, Anna worked in education in South Africa and the UK. When she returned to the states, she enrolled in graduate school at Columbia University on a path to work in public education and economic development. Following a pivotal conversation with her mother about the possibility of Frances writing a sequel to her book, *Diet for a Small Planet*, the two decided to team up and travel the world together to do field research, seeking out stories of social movements addressing the root causes of hunger.

Anna traveled to Poland alone, as well as to India, Bangladesh, Kenya, France, Brazil, and around the U.S. with her mother. They collected numerous research notes for the book. Over the course of a year, their research began to tell a story culminating in the book *Hope's Edge*. Anna says she was, "very moved by that experience."

In India, on their way to meet a network of farmers promoting organic agriculture, Anna saw a grove of beautiful eucalyptus trees. Painted across all of them were giant Pepsi logos.

"It was ... easier for villagers to get Pepsi from their local distributor than it was for them to have access to clean drinking water." Anna explained that there were dozens of examples just like that one that portrayed how much "U.S. foreign policy or the U.S. corporations were impacting farmers and eaters a world away." Moreover, she saw how eating American processed food was seen as "a point of pride" for many people around the world. Anna realized the full extent of how much U.S. food and soda companies' marketing and political campaigns were impacting the health of communities across the globe.

Anna at the International Federation of Organic Agricultural Movements conference in Istanbul, Turkey

"Start thinking of food through the lens of sustainability, as well as justice, and pay attention to who has access to healthy food, as well as fairness to farmers."

"She just did what she felt needed to be done and took it one step at a time."

Maathai began building a village-based women's movement to grow a massive tree planting initiative. She advocated for training this female workforce, what she dubbed "foresters without degrees." Tens of millions of trees were planted.

"In the process of the tree planting work, which was so masterful, she listened to what village women expressed were their needs," says Anna. This included protecting the soil, creating food crops, and re-greening their landscapes. Trees can meet these needs. "I remember thinking we had just met the most amazing person ... and a few years later, she decided to run for student parliament and won by a landslide. Then, in 2004, she won the Nobel Peace Prize for her work. I think about what was so amazing about her story. It's that she never sought out to do it or get this enormous international acclaim. She just did what she felt needed to be done and took it one step at a time."

"We were well aware of how traditional Brazilian diets are so healthy and yet saw how much advertising there was from American food brands," she says, describing their time in the Brazilian city of Belo Horizonte "and it struck me ... high rates of childhood malnutrition, high rates of poverty, and the government of Belo Horizonte was having to spend hard-earned tax dollars on public education as a counterweight to the massive amount of marketing that McDonald's and soda companies were pouring into their city and it was so crazy and infuriating to me."

During this time, Anna and her mother met the late environmental and political activist Wangari Maathai, the founder of the Green Belt Movement in Kenya. Maathai saw the desertification of Kenya happening right before her eyes. Without missing a beat, she did what needed to be done.

Maathai empowered women to trust themselves and their leadership. As a result, there was a massive shift of ordinary people realizing their power. Wangari Maathai had a profound impact on Anna and her future work. Anna became passionate about influencing how food policies are shaped and how the U.S. thinks about food systems. She has not looked back since, and she works tirelessly to promote food sustainability and fair food systems.

Dear Wangari,

I was just listening to a recording of you recounting one of the stories you loved to tell: Once upon a time, there was a fire in the forest. Faced with the towering flames, a hummingbird was rushing between a nearby lake and the tall trees, bringing water back and forth, back and forth. Its tiny beak was gathering drops of water to combat the terrifying blaze. The other animals in the forest scoffed at the colorful bird: *What are you doing*, they asked? *What difference can you make?* The hummingbird responded: *I'm doing what I can.* Which is what any of us can do—we do what we can. Hearing your voice, I could see the places you took us to in Kenya like it was just yesterday: The village trees, the kitchen gardens, the incredible women leaders in the Green Belt Movement, and the dancing—always the dancing.

Thank you for showing the world the intimate connections between caring for the environment, nourishing ourselves through food, and promoting peace and democracy. Your spirit lives on in the countless lives you touched, including mine.

Always,

anna

ANNA LAPPÉ'S RECIPE:

Wendy's maple amaranth porridge

prep time: 10 minutes • *cook time*: 25 minutes • *makes*: 3-4 servings

This nourishing recipe is the ultimate cold-morning comfort food. It was shared with me by the wonderful Wendy Lopez, a nutritionist, cookbook author, and co-founder of Food Heaven Made Easy.

When I set out to revamp the vegetarian recipes of my mother's 1971 classic, *Diet for a Small Planet*, for the book's 50th anniversary (goodbye soy grits and margarine!), I turned to Wendy. She has a gift for simple nourishing recipes that bring to life the beauty, textures, and flavors of plant-centered eating. Together, we updated some of the 100+ recipes in the original and invited more than a dozen of our favorite chefs and culinary maestros to contribute their own, too. Along the way, I came to appreciate Wendy's eye for healthy and delicious eating and now, when I make myself this porridge, I think about our many Zooms and cross-country chats about measurements and spices—and all things delicious. For my version of this recipe, I love topping with sliced fresh figs, chopped almonds, and a dusting of brown sugar. *anna*

ingredients

1 cup amaranth

¾ cup unsweetened almond milk or other plant-based milk, warmed

1–1½ tablespoons maple syrup, to taste

¼ teaspoon vanilla extract

½ teaspoon ground cinnamon

2 teaspoons coconut oil

brown sugar, figs, blueberries, sliced almonds, pumpkin seeds, and/or dried cranberries for optional toppings

Adapted by Anna Lappé from Wendy Lopez's Maple Amaranth Porridge in Diet for a Small Planet: The 50th Anniversary Edition *by Frances Moore Lappé (Random House 2021)*

directions

1. In a medium pot, bring 3 cups water to a boil and add the amaranth. Turn down heat to simmer, cover, and cook, stirring occasionally, for 20 minutes, or until most of the water has been absorbed.

2. While it's cooking, prep any of the toppings you'd like to enjoy: slice figs, chop nuts, dice fruit.

3. Once done, remove from stove and add the almond milk, maple syrup, vanilla, cinnamon, and coconut oil. Stir well.

4. Ladle the mixture into bowls and top with your favorite dried fruits, nuts, and/or seeds. You can store leftovers in the fridge for up to a week. When it's time to reheat, add additional almond milk, so the porridge isn't so thick.

ocean + fisheries

Growing up on a small dot in the middle of the Pacific Ocean, you learn to be mindful of all that surrounds you, which in the most literal sense, is water. Ancient Hawaiians were water people. By studying the tides and the moon and noticing that fish would gather in certain areas, Hawaiians came up with an ingenious way to collect fish by creating loko i'a, or fishponds, by closing off a section of the ocean to farm fish in their natural habitat. My ancestors made use of natural resources in an incredible way, understanding that if they cared for the land and sea, it would take care of them. A lot of people may not realize that modern-day fishing and fish farming have changed tremendously since then, and we felt it was important to share this. *mackenzie*

Most of what you have read in this book focuses on land-based sustainability. While this is vitally important, we cannot forget about our *oceans*. Everything is connected, and just like what we consume on land, everything we consume from the oceans has an impact on our health and on the environment. In this fact sheet, we share some major challenges facing our marine life today.

Overfishing and industrial fish farms detrimentally impact marine life and communities that depend on the fishing industry. Overfishing occurs when a species, like tuna or salmon, is caught at a rate that rapidly overpowers their ability to reproduce. In half a century, the number of fish that have been overfished has tripled. This rate has increased exponentially due to bycatch, another major threat to our oceans. Bycatch is the capturing of unwanted sea life while fishing for a separate species, which ultimately leads to the avoidable deaths of billions of fish, sea turtles, and crustaceans. For example, in 2007, the Alaska pollock industry accidentally caught 120,000 king salmon in pollock nets. By law, these now-dead fish were thrown overboard. These practices of overfishing and bycatch result in imbalanced ecosystems and ultimately cause the loss of other marine life that can no longer feed on their typical prey.

Industrial fish farms, which produce about half of the fish eaten around the world, are another factor harmful to marine life. As of 2012, more farmed fish than beef are produced globally. In these environments, fish are kept in unnatural enclosed areas. These crowded enclosures cause fin damage and promote disease outbreaks that are ultimately treated with pesticides. In fish farming, chemicals like PCBs and antibiotics are put in the fish feed and cause immense damage to wildlife, through waterway leakage. Most farmed fish, such as salmon, therefore contain these PCBs, an extremely toxic chemical that is also used in computer chips and flame retardants. Additionally, PCB concentrations in farm-raised salmon were almost 8x higher than that of wild salmon, and frequent consumers of farmed salmon may exceed government health limits for these pollutants, which are linked to immune system damage, fetal brain damage, and cancer.

See page 230 for a list of resources for how you can get involved and help!

Bycatch ultimately leads to the avoidable deaths of billions of fish, sea turtles, and crustaceans.

smoked tomato "lox" with gluten-free bagels

prep time: 15 minutes • **cooling time:** 30 minutes • **serves:** 6

The ethics of eating lox (smoked salmon) in today's world is complicated. Not only do salmon play an important role in our ecosystem, but they are seen as a vital food source and a relative of Indigenous peoples who live in places such as the Pacific Northwest and Alaska. But today, Pacific salmon are extinct in 40% of the rivers of Washington, California, Oregon, and Idaho. Due to the increasing demand for human consumption, approximately 70% of salmon produced worldwide is farmed, and the majority of farmed salmon are contaminated with PCBs (industrial chemicals banned by the EPA in 1979). I know ... depressing news, and trust us, as enthusiastic Jewish eaters, we desire nothing more than a good bagel and lox. This is why we were thrilled to discover that a touch of kelp powder and liquid smoke will have you believing you are eating real fish, without the negative impact of the industrial fishing industry and the byproduct of PCBs.

ingredients

6 large Roma tomatoes

2 teaspoons liquid aminos or tamari

2 teaspoons olive oil

1 teaspoon liquid smoke

1 teaspoon water

1½ teaspoons kelp powder

topping ideas

capers • red onion

*Liquid smoke gives you the smoky flavor without the hassle of using a smoker! It even comes in a variety of flavors such as hickory and mesquite to mimic the many types of wood used for smoking.

directions

1. Bring a medium saucepan filled with water to a boil. Pierce the skin of each tomato with the tip of a paring knife, then drop the tomatoes into the boiling water to blanch. Boil for 1 minute.

2. Remove tomatoes and place them in a bowl of cold water.

3. Combine tamari (or liquid aminos), olive oil, liquid smoke, water, and kelp powder in a small bowl. Whisk until mixed.

4. Slip the skins off the tomatoes and compost. Cut tomatoes in half, trim away the seeds from the firm inside layer, and drop tomato halves into the bowl of tamari mixture.

5. Mix well to ensure tomato halves are all evenly coated with marinade.

6. Chill in the fridge for 30 minutes before serving. Enjoy your tomato "lox"on toasted bagels (page 40) with suggested toppings.

*With the inclusion of kelp powder,
there is still a taste of the ocean without the presence of the fishing industry. Plus, it includes minerals such as potassium, calcium, and magnesium.

Frappé

pre-prep: 15 minutes to brew coffee or or use coffee that's been refrigerated overnight

prep time: 15 minutes • **before serving**: Place in freezer for 30 minutes (optional) • **serves**: 2

I must confess, I've never had a Frappuccino. They had too much sugar for my mom's approval, but I was always curious about what they tasted like! I know a lot of people get their daily fix from Starbucks, so we thought it would be fun and important to come up with an alternative that has the same delicious taste and caffeine fix. *mackenzie*

ingredients

1 banana, fresh or frozen

1 cup coffee, freshly brewed or cold brew

1 cup almond milk

⅓ cup coconut milk

2 dates, pitted

1 teaspoon maple syrup (*optional, depending on desired sweetness*)

1 scoop chocolate protein powder (*about 3 tablespoons*)

½ teaspoon cinnamon

topping ideas

coconut whipped cream
We recommend the So Delicious CocoWhip.

chocolate shavings
Chocolate Sauce (*page 214*)

directions

1. Add all ingredients into a blender and blend until smooth.
2. Pour into two cups and place in freezer for 30 minutes.

* We like to **make the coffee the night before** and pour it into large ice cube trays and freeze! In the morning, pop out the cubes for a colder drink.

gluten-free bagels

prep time: 25 minutes • *bake time*: 25 minutes • *makes*: 6 bagels

It's difficult to find a fresh gluten-free bagel, let alone one that is not dense as a brick. Coming up with this recipe was worth all of our bagel failures. Not only are these bagels light and chewy, but the addition of yogurt gives the bagels a sourdough-like taste. Be creative and try this recipe with different yogurt flavors, such as blueberry, strawberry, or vanilla. If you're hoping for some tasty toppings, check out our spread suggestions under the recipe's topping ideas.

ingredients

1 cup gluten-free flour (*keep ½ cup extra flour on the side for dusting and adding when needed*)

2 teaspoons baking powder

½ teaspoon salt

1 cup plain plant-based yogurt

1 flax egg (*page 10*)

small bowl of lukewarm water with a dash of olive oil for hands

seasoning ideas

sesame seeds • salt • pepper
garlic powder • onion powder

topping ideas

Almond Tofu Ricotta Cheese (*page 182*)
plant-based cream cheese
Smoked Tomato "Lox" (*page 36*)
Vegan Fried Egg (*page 177*)

directions

1. Preheat oven to 400°F. Line a baking sheet with parchment paper.
2. In a large bowl, combine flour, baking powder, and salt. Whisk to combine.
3. Add yogurt and flax egg and stir to combine.
4. Dip your clean hands into olive oil and water mixture to lessen the stickiness of the dough.
5. Lift dough out of bowl and onto a clean surface (such as a cutting board) dusted with extra flour. Knead dough with oiled hands, adding more flour as needed.
6. Separate the dough into 6 equal portions. Roll each into a ball between your cupped palms. Press each round of dough into a disk about ½ inch tall. With your floured finger, create a 1-inch hole in the center of each disc. Place each bagel on the parchment paper–lined baking sheet.
7. Lightly coat each bagel with seasonings of your choice. Bake for 22–25 minutes. Let cool before slicing.

hemp granola

prep time: 15 minutes • cook time: 1 hour • makes: 20 servings

When it comes to granola, it is definitely worth it to make your own. Most store-bought granola contains a lot of sugar and highly processed oils. Our Hemp Granola recipe is loaded with nutrient-dense nuts and seeds, which are full of antioxidants and lightly sweetened with maple syrup. After seeing how easy and delicious this is, you'll never want store-bought granola again!

ingredients

¾ cup almonds

½ cup pumpkin seeds

1 cup unsweetened coconut flakes

½ cup pecans

1 cup walnuts

1 cup old-fashioned oats

1 cup hemp hearts

2 tablespoons chia seeds

⅓ cup coconut oil, melted

1 teaspoon vanilla extract

1 teaspoon cinnamon

¼ cup maple syrup

pinch of salt

directions

1. Preheat oven to 250°F. Line 2 baking sheets with parchment paper. Place almonds, pumpkin seeds, coconut flakes, pecans, and walnuts into a food processor. Pulse about 10 times until mixture is chopped but still a good, chunky texture.

2. Transfer nut mixture into a large bowl. Add all remaining ingredients and stir well. Spread an even layer of the mixture onto each baking sheet.

3. Bake for 35 minutes. Remove from oven and give the mix a good stir. Bake for an additional 25–30 minutes more until granola is golden brown.

4. Remove from oven and let cool. It will get crunchier as it cools. Store in mason jars in the refrigerator to keep it extra crunchy.

*Hemp hearts are loaded with protein and essential fatty acids, omega-3 and omega-6. They also contain magnesium, fiber, and iron.

bob MOORE

Bob Moore is the founder and president of Bob's Red Mill Natural Foods, which manufactures and sells a variety of whole grain products including stone-ground flours, cereals, meals, and more.

Bob founded the company in 1978 with his wife, Charlee, based on their shared commitment to providing nutritional whole grains that support a healthy lifestyle. In 2010, Bob announced to his employees that an Employee Stock Ownership Plan (ESOP) would be put into place. This government-approved program gave each employee a share in the ownership of his company and, according to Bob, the ESOP, "allowed the continuity of the company to be secured." Bob did not retire, but is now himself a member of the ESOP. The company is still dedicated to supplying grocery stores across our country, and around the world, with freshly stoned-milled products. Bob's Red Mill was voted the #1 Most Admired Company in Oregon for eight straight years from 2011 through 2018. You may very well find Bob's face looking at you from his products already in your own cupboards!

Bob Moore likes to say that although he got a lot of inspiration from books, his wife Charlee was the real reason behind the creation of Bob's Red Mill Natural Foods. This story is about their partnership, but also about defying the odds and finding the strength to rebuild after their 1988 mill fire.

Bob comes from a family with deep roots in the food business. His great-uncle Charles was president of the Continental Baking Company, and his great grandfather had a food business of his own. Charlee's grandmother, Donno, was fascinated with Adelle Davis, a health food book author and one of America's best-known nutritionists during the 1940s–1960s. Her books, and others, influenced Charlee and Bob when they and their three sons moved to a five-acre dairy goat farm in Northern California. Charlee used Donno's collection of books to create a healthier lifestyle for her own family. As they raised their sons, the importance of whole grains became abundantly clear to the Moores. With the added bonus of Bob's lifelong love of machinery, the Moore family started thinking about setting up a mill of their own.

a bit of history

For millennia, the world's grain foods were milled by hand between two stones. The stone-ground flour was by necessity whole grain—rich in fiber, germ, and nutrients. Around the year 1880, high-speed steel roller mills replaced the slow-turning stone mills that had supplied the world with flour for centuries. The new machines separated the bran and germ from the white endosperm, unlike the slow-turning stone mills that preserved all parts of the whole grain.

In **1968**, the Moore family's life changed when Bob walked into a public library and saw the book *John Goffe's Mill* by George Woodbury laying on a table. It's the story of an archeologist who inherited and restored a crumbling gristmill and began supplying wholesome whole grain flours and meals to stores in his rural New Hampshire area. Bob says the story quickly became the inspiration of his life. "If he could do it, so could I!"

In **1969** and **1972**, Bob and his family visited many old flour mills around the United States and Canada searching for usable millstones from the **1880s**. With the help of a miller friend, Bob was able to acquire several sets of stones from an old water-powered flour mill in North Carolina. This allowed him to set up his first operation, Moores' Flour Mill, in Redding, California with Charlee and two of their three sons. The undertaking prospered and, in fact, their sons are still operating it there.

In **1978**, Bob and Charlee returned to Bob's birthplace in Portland, Oregon where he planned to attend seminary. But, soon after Bob enrolled in seminary to study Biblical Hebrew and Greek, the couple happened upon an abandoned mill near the school. When Bob was 49, he and Charlee plunged into business again. They bought the mill, painted it red, and salvaged several sets of 150-year-old, 2,000-pound quartz millstones from an abandoned mill in eastern Oregon. From here, Bob's Red Mill was born.

In **1988**, an arsonist burned down their historic gristmill, destroying everything except the stone mills. During the fire, hundreds of pounds of stored grain fell from the second floor and covered the millstones below. The grain insulated the stones from being destroyed by the fire, as well as preserved the gears that turned them.

The Moores never gave up and set their sights on rebuilding Bob's Red Mill. Today, there is a good, steady market for whole grains, and Bob offers 400-plus products in over 40 countries across the globe. In his biography, *People before Profit*, Bob describes the brand in his own words: "It's an authentic brand, with more than 750 *real* people grinding millions of pounds of *real* whole grains, mixing them with *real* ingredients, and packaging them in eco-friendly stand-up pouches with 'windows' so that customers can see the quality of our products." Bob's face is on every bag, his personal commitment to the excellence of his company's products.

In *People before Profit* Bob says, "I envisioned the mill as a change in my life to be able to do what I wanted to do and what I believed in, which was whole grains ... exposing 150-year-old machinery to the public so they could appreciate what I was doing. It was unique, it was healthy, and it fit all my aspirations for helping people."

Bob Moore is someone who has spent his time pursuing an out-of-the-ordinary way of life. First, he explored the all-but-lost craft of stone milling before whole grain foods were trendy, and then he set about restoring his fire-ravaged business at age 59, when many people start thinking about retirement.

Over the years, Bob had many opportunities to sell his company, but this was of no interest to him. In **2010**, on the occasion of his 81st birthday, Bob and his partners signed the papers to create an ESOP to gradually give the company to the folks who worked for him and helped make Bob's Red Mill a real success. In 2020, after ten successful years as an ESOP, the company completed the transfer of 100% of Bob's Red Mill's stock to its employees.

2022: Bob, at 93, is energized by the work he does. You will find him every day at the company's headquarters, and he still works full time as the company's founder and president. Bob Moore is an inspiring example of someone who loves what he does and is enjoying the benefits of good health and longevity that can come from a whole-grain diet, regular exercise, and a keen interest in life.

Sweetheart,

I miss you a great deal! As I review our life's success together, I thank you profoundly for your motherhood, your wifehood, and your advice and guidance that inspired our wonderful life together making whole grains and distributing them worldwide!

Thank you,

bob

BOB MOORE'S
baked pear oatmeal
from bob's red mill

prep time: 15 minutes • **cook time:** 40–50 minutes • **makes:** 8 servings

ingredients

2 cups Bob's Red Mill Old Fashioned Rolled Oats
   or Organic Regular Rolled Oats

1 teaspoon baking powder

½ teaspoon salt

½ teaspoon ground cinnamon

¼ teaspoon ground ginger

¼ teaspoon ground nutmeg

3 cups water or milk, pear juice, or apple juice

2 eggs

½ cup brown sugar

¼ cup butter, melted and cooled

1 teaspoon vanilla extract

2 cups chopped fresh pears *(about 2 whole)*

directions

1. Preheat oven to 350°F. Grease a 9x9" baking dish with butter or coconut oil and set aside.

2. In a small bowl, combine rolled oats, baking powder, salt, and spices. Set aside.

3. In a large bowl, whisk together water (or milk or juice), eggs, brown sugar, melted and cooled butter, and vanilla extract. Add oat mixture to the wet ingredients and mix to combine. Fold in chopped pears.

4. Transfer mixture to the prepared baking dish and bake until set, about 40–50 minutes.

5. Let cool about 5 minutes before serving.

To make it plant-based:
- plant-based milk can be substituted for milk
- flax eggs can be substituted for eggs (page 10)
- vegan butter can be substituted for butter

Bob offers 400-plus products in over 40 countries across the globe.

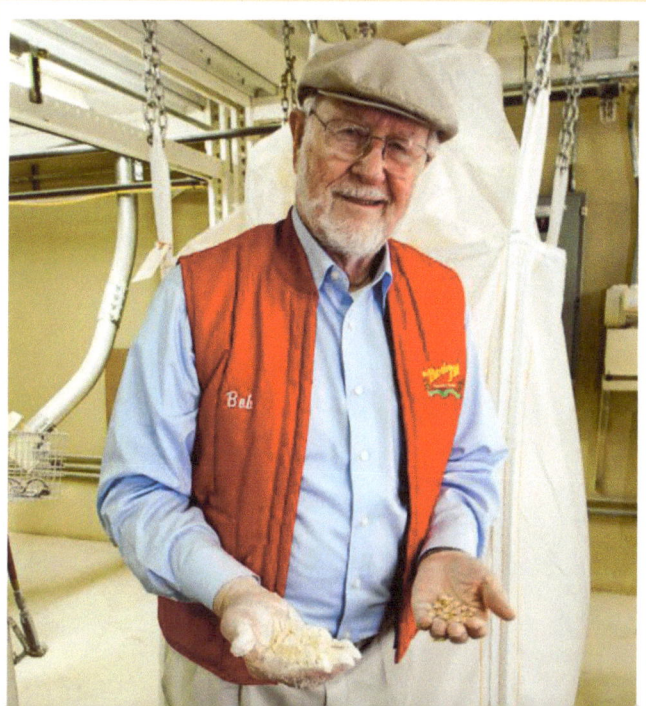

peanut butter & jelly overnight oats

prep time: 25 minutes • *before serving: Keep in refrigerator overnight* • *serves: 2*

We've been fans of the idea of overnight oats for a while—with just simple prep the night before and voilà, a filling breakfast you can take on the go. It took some trial and error to come up with the right balance of flavors, but we finally created an original that tastes like no other. It's packed with fiber and protein, and it's so flavorful you'll want to jump out of bed in the morning and head straight to the fridge!

ingredients

oats mixture

1 ripe banana

2 cups plant-based milk

1 tablespoon maple syrup

¾ cup oats

1 teaspoon cinnamon

1 tablespoon chia seeds

1 tablespoon ground flaxseed

1 tablespoon hemp seeds

4 tablespoons peanut butter powder

2 teaspoons peanut butter

topping mixture

8 strawberries, fresh or frozen

1 banana

¼ cup plant-based vanilla yogurt

directions

1. In a medium bowl, mash banana with a fork. Add nut milk and maple syrup and mix.
2. In another medium bowl, combine oats, cinnamon, peanut butter powder, plus chia, flax, and hemp seeds. Using a fork, mix well.
3. Pour banana mixture into oat mixture and mix well. Divide mixture into two separate cups. In each cup, add 1 teaspoon peanut butter and mix.
4. Cover cups and store in refrigerator overnight.
5. Place strawberries, banana, and yogurt into a blender and blend well.
6. Store mixture in refrigerator and pour over oats in the morning.

slow cooker oatcakes

pre-prep: 15 minutes to cook quinoa, 15 minutes to soak seeds and nuts • **prep time**: 20 minutes
cook time: 2½ hours • **makes**: 12 Oatcakes

This is one of our most essential recipes, as we have an oatcake every single morning! The combination of oats, quinoa, nuts, and seeds is packed with fiber and protein, keeping you full and energized until lunch and beyond. One loaf provides you with a grab-and-go breakfast for the next 12 days!

dry ingredients

2 tablespoons chopped walnuts

1 tablespoon hemp seeds

1 tablespoon chia seeds

1 tablespoon ground flaxseed

1 tablespoon pumpkin seeds

1 cup rolled oats

¼ cup steel cut oats

1 tablespoon peanut butter powder

1 teaspoon cinnamon

2 tablespoons dark chocolate chips

1 cup cooked quinoa

wet ingredients

¼ cup maple syrup

¼ cup coconut oil

¼ cup almond butter

4 bananas

toppings

¼ cup coconut flakes • 1 tablespoon cacao nibs
1 tablespoon dark chocolate chips

directions

1. Combine walnuts, hemp, chia, flax, and pumpkin seeds in a small bowl. Add ¼ cup water and soak for 15 minutes.

2. In a large bowl, add rolled oats, steel cut oats, peanut butter powder, cinnamon, and chocolate chips. Mix well.

3. Add maple syrup, coconut oil, almond butter, and bananas to a blender and blend until smooth.

4. Add blender mix and seed mix to dry mix and stir.

5. Add cooked quinoa.

6. Pour ingredients into slow cooker lined with parchment paper.

7. Top mixture with coconut flakes, cacao nibs, and chocolate chips.

8. Cover and cook in the slow cooker on low for 2½ hours.

9. Remove from slow cooker and keep the loaf in the parchment until cooled.

10. Makes 10–12 oatcakes.

11. Cut into squares and serve. Refrigerate or freeze the rest.

Peanut butter powder is high in protein and rich in fiber, but has 85% less fat than regular peanut butter. *Oats* are a good source of soluble fiber.

DR.
malia
SMITH

Malia picking ulu (breadfruit) from her tree.

Dr. Malia Smith is the owner of the plant-based restaurant and wellness center 'Ai Love Nalo in Waimānalo, Hawai'i.

Smith received her doctorate degree in education from the University of Southern California (USC), and her scholarly work revolves around cultural impacts upon behavior and self-efficacy. She is also the president and CEO of Sustainable Ideation, LLC, which has been contracted to provide sustainability solutions, research, and reports for groups including the Office of Hawaiian Affairs and the Hawai'i Department of Agriculture.

DR. MALIA SMITH'S
ulu (breadfruit) waffles

prep time: 10 minutes • **cook time:** 30 minutes • **makes:** 6–8 waffles

ingredients

½ cup cashew milk

2 teaspoons vanilla extract

1 tablespoon apple cider vinegar

2 flax eggs *(page 10)*

1½ cups ulu flour or gluten-free flour

 Ulu flour can be purchased at Hawaiianfarmersmarket.com

½ tablespoon baking powder

½ teaspoon salt

3 tablespoons coconut sugar or maple syrup

5 tablespoons vegan butter, melted

topping suggestions

sliced bananas and berries

directions

1. Preheat waffle maker. In a small bowl, mix cashew milk, vanilla, and apple cider vinegar. Set aside for 5 minutes to make "buttermilk."

2. In another bowl, mix flax or chia meal with water to make vegan eggs. Set aside.

3. In a medium bowl, mix flour, baking powder, salt, and sugar.

4. Add the melted butter and egg to the buttermilk mixture and mix well.

5. Slowly add wet mixture to the medium bowl that contains the dry ingredients and mix the wet and dry ingredients together.

6. Add a large scoop of mixture to the waffle maker. Cook until ready and top with bananas and berries.

WHAT IT MEANS TO BE HAWAIIAN

RESPECT

Born and raised on the fertile grounds of Nanakuli, Hawai'i, Dr. Malia Smith and her family embodied "sustainable living" before it was popular. Eating off the land was a natural existence for her family and neighbors. Smith learned how to respect and co-exist with the land at a young age, living by the concept of "mālama 'āina" (to care for the land). Despite her family's lower socioeconomic standing, their respect for the land produced abundance: "It was something that was natural to me. It made sense, and it was a way of life ... So, interestingly enough, growing up low-income, I did not feel deprived, because we lived off the land."

NURTURE

After receiving her doctorate degree from USC, Smith returned to Hawai'i to nurture her community. It was an awareness of the community's increased dependence on fast food and the relationship between this type of diet and illness that prompted her to bring farm-fresh ingredients back to the people of Hawai'i. In 2015, Malia opened 'Ai Love Nalo, a plant-based restaurant and wellness center in Waimāna-lo, where there is a predominantly Native Hawaiian population.

"Our deep-rooted commit- ment is to this community."

Smith aims to make healthy food accessible to everyone. 'Ai Love Nalo serves locally sourced organic meals that are healthy and tasty, while sup- porting local farmers who practice sustainable farming methods.

"Growing up low-income, I did not feel deprived, because we lived off the land."

As part of the center's mission, Smith uses food to address health problems. In 2019, she started the Hā'ehuola program, which provides Native Hawaiian participants with the tools to lose weight. She offers plant-based meals for 12 weeks and over 50 hours of classes in nutrition, cooking, shopping, and gardening to help people transition to a healthier lifestyle.

GIVE

We are part of a larger purpose. When asked about 'Ai Love Nalo's role in the community, Smith explained that her generosity is about much more than herself: "We have a saying, **'na wai ke kupu o oe,'** which means 'whose seed are you?' This Hawaiian proverb means a lot to us because the deep meaning behind it is: 'Who do you come with? Who do you represent?'

I taught this to my daughter when she was very young. For us, it means, when you are engaged in something, you are not standing alone by yourself. You are bringing all of your ancestors with you. You represent all of those who came before you, as well as all of the people that are coming after you.

It also means that we must do things from the heart, because our lives, and everything we have, is on loan—there is no ownership. 'Ai Love Nalo is under our name, but **the reason we're here is for the community that we represent ... We are part of a larger purpose."**

Dear Gramma,

Thank you for teaching me what it means to be Hawaiian. Through you, I have come to know that all things are connected, and that life is a beautiful cycle. You taught me that it all comes down to five simple, yet powerful words: Respect. Nurture. Give. Receive. Appreciate. Thank you for being one of my greatest teachers!

With Love,

malia

RECEIVE

'Ai Love Nalo is a place that both gives and receives. Smith receives local produce from organic farmers in Hawai'i and provides fresh food to her community. The origin of this name speaks to its mission:

"Simply stated, we believe that 'Ai (to eat) Love means food that we provide with love. Nalo, short for Waimānalo, where our restaurant is located, means potable water, and to the Hawaiian people, water represents life.

By saying, "I love Nalo," we represent the aloha we all share. It represents the loving food and fresh water one can enjoy here, in Waimānalo, at 'Ai Love Nalo restaurant."

APPRECIATE

Smith's mission goes far beyond her restaurant. Her appreciation for the land is marked by her commitment to sustainability, in every sense.

"To me, sustainability is much more than just hugging trees and taking care of the environment. It's all of it. It's financial sustainability, it's your body, it's every aspect."

By looking to the past, as well as the future, and by taking care of the present, Smith embodies Hawaiian epistemology.

We appreciate Smith for doing this work for our community.

GMOS + monoculture

We've all heard about GMOs, but do we really know why it's better to choose non-GMO? A "genetically modified organism" is defined as a plant, animal, microorganism, or other organism whose genetic makeup has been modified in a laboratory using genetic engineering or transgenic technology. This process creates combinations of plant, animal, bacterial, and virus genes that do not occur in nature or through traditional cross-breeding methods.

Among the problems caused by GMOs, they've led to a situation where just four companies (Bayer, Corteva, ChemChina, and BASF) control over 60% of the seed market. How did this happen? Genetically engineered (GE) varieties of seed were introduced in 1996, and by 1998, the large agribusiness companies had accelerated their consolidation by buying up smaller firms. By 2008, Monsanto's (now Bayer) patented genetics alone were planted on 80% of U.S. corn acres, 86% of cotton acres, and 92% of soybean acres. Today, these percentages are even higher.

GMOs have led to a situation where just four companies control over 60% of the seed market.

This concentration has made a huge dent in farmers' pockets. USDA data shows that the per-acre cost of soybean and corn seed spiked dramatically between 1995 and 2014. Those costs far outpaced the market price farmers received for corn and soy, leaving them with tighter margins on which to run their farms. Farmers who buy GMO seeds must also pay licensing fees, sign contracts that dictate how they can grow the crop, and even allow seed companies to inspect their farms. These companies also restrict research on seeds that farmers or independent researchers may want to conduct.

Counter to what many people believe, there are only a handful of GM crops: primarily corn, cotton, and soy. These GM crops are often broken down into two categories: herbicide tolerant and plant-incorporated protectants (PIPs), meaning the crop is either designed to be made resistant to the herbicide, or the crop is designed to produce the pesticide internally. Herbicide-tolerant crops are designed to tolerate specific broad-spectrum herbicides, which kill the surrounding weeds, but leave the cultivated crop intact. Today, about 94% of soybeans and 89% of corn grown in the U.S. are herbicide resistant. This encourages farmers to freely apply more herbicides, which will destroy any surrounding weeds but won't kill crops grown from GM herbicide-resistant seeds. This excessive herbicide use has led to herbicide-resistant weeds, known as superweeds, as well as superpests that are extraordinarily difficult for farmers to manage.

Another issue with GMOs is the contamination and resulting economic loss. Because plants are pollinated by insects, birds, or wind, pollen from a GMO plant can move into neighboring non-GMO fields, contaminating farms and causing organic farmers to lose their organic certification and face rejection from export markets that ban GMOs. Seed corporations can even sue farmers whose farms get contaminated for "seed piracy." (Watch the film *Percy vs. Goliath* to learn a true story about this.)

GMOs also promote monoculture, or the cultivation of a single crop in an area. Why does this happen? There are only a handful of GM crops, and the majority are made to be resistant to herbicides, such as Bayer's Roundup Ready 2Yield® Soybeans and Roundup Ready® Corn2. This reality leaves farmers without the option of growing a healthy variety of crops because the herbicides used on the GM crops would destroy anything else farmers might want to grow. In this way, GMOs threaten the diversity of our food supply (think of the Irish potato famine, when a disease destroyed the potato crop and there weren't enough other crops to feed the population). In addition, fields with genetically modified crops are not only mono-crops, but are also genetically identical. This means that there is no variety of nutrients to enrich the soil.

Monoculture also decreases biodiversity and throws the ecosystem out of balance. For example, insects inhabiting a monocultural farm may not have predators, and their populations can grow out of control. In an organic environment with a diverse collection of plants and animals, each has a unique purpose, and genetic diversity helps species acclimate to new pests, diseases, and environmental conditions like droughts. Tools like traditional breeding techniques and seed banks, which preserve the genetic diversity of seeds, are proving effective at developing drought-tolerant crops.

so what else can we do?

For individuals, saving seeds is a beautiful act of peaceful resistance and community building. Saving seeds is not only self-sustaining, but it also helps take action against major corporations like Bayer who are trying to sue farmers who don't abide by their rules, and dominate the seed industry. It is crucial to support local farmers who are growing food without using harmful herbicides and pesticides. Small farmers can be found at your local farmers' market, where you can chat with them about their practices. You can also find out which farms offer Community-Supported Agriculture (CSA) programs, where they fill a box with freshly harvested produce every week for a set price, and whether they can either deliver it directly to you or have a pick-up spot in your area.

But individuals acting alone isn't enough: Structural, collective fixes are needed, like more state- and national-level support to start or transition to biodiverse, regenerative organic farms. Farmers also need affordable access to seeds, the right to save them, and freedom to conduct seed research.

main dishes

chickpea "crab" cakes

pre-prep: If using dried chickpeas, soak for 6 hours or overnight and cook for 1 hour (page 11)

prep time: 25 minutes • *cook time*: 15 minutes • *serves*: 4

Who knew that heart of palm has the potential to transform into an amazing "crab-like" taste and texture? By lightly shredding and seasoning with kelp powder and an array of spices, you really will feel like you're munching on crab cakes. Plus, you get to feel good that you left the crabs alone to play the essential role they have in our marine ecosystem and did your part to prevent overfishing and bycatching in the process. Enjoy these patties as an appetizer, side, or main dish.

ingredients

1 cup chickpeas, cooked or canned, drained
   From either method, set aside 2 tablespoons chickpea liquid.

1 can (*14 ounces*) hearts of palm,
   drained, sliced to fit in food processor

½ cup gluten-free breadcrumbs

1 teaspoon onion powder

1 teaspoon garlic powder

1 teaspoon kelp powder

1½ teaspoons Old Bay Seasoning

½ teaspoon dried parsley

½ teaspoon salt

dash of pepper

2 tablespoons vegan mayonnaise
   Check the ingredients! We prefer a brand without canola oil such as Chosen Foods Vegan Mayo.

2 teaspoons lemon juice

2 tablespoons olive oil

½ teaspoon Worcestershire sauce

1 teaspoon Dijon mustard

2 tablespoons onion, finely chopped

3 cloves garlic, minced

¼ cup gluten-free breadcrumbs for coating patties

2 tablespoons olive oil for frying

serving ideas

rice of choice • vegan tartar sauce • Roasted Bell Peppers (*page 152*)

We love brown basmati rice!   *There are plenty of easy recipes online!*

* **Heart of palm** is a good source of iron, potassium, copper, phosphorus, and zinc.

directions

1. In a food processor, add the chickpeas and hearts of palm together. Pulse just a few times to break up. Don't pulse too many times or it will lose its crab-like texture. Set aside.

2. In a large mixing bowl, add breadcrumbs, onion powder, garlic powder, kelp powder, Old Bay Seasoning, dried parsley, salt, and pepper. Stir all dry ingredients to combine.

3. In a small bowl, whisk chickpea liquid until slightly foamy. Add mayonnaise, lemon juice, olive oil, Worcestershire sauce and Dijon mustard.

4. Add chickpea "crab" and mayonnaise mixture to large bowl, along with onion and garlic cloves. Mix all ingredients together.

5. Place breadcrumbs for coating on a shallow plate. Form small patties with hands and coat evenly with breadcrumbs. Set patties aside on a large plate.

6. Heat large skillet over medium heat. Add 1 tablespoon olive oil and wait until the pan is sizzling before adding patties. Pan-fry the patties for about 3–4 minutes on each side until golden brown. Flip each patty once, then remove from heat and add 1 more tablespoon olive oil to finish cooking.

7. Serve immediately with suggested sides.

land access

To mitigate climate change in the agriculture sector, we cannot focus on farming practices without first addressing who has access to farmland. We must lower barriers to entry for new farmers who are interested in practicing regenerative, organic agriculture. And, we must rectify historical disparities when it comes to land access and tenure for BIPOC (Black, Indigenous, and People of Color) farmers, who have faced a horrific history of discrimination in credit markets, state and federal farm programs, and real estate. Despite all of the barriers BIPOC farmers have faced in order to farm in this country, agriculturalists of color have made tremendous contributions when it comes to sustainable farming. Learn about some of these folks in Groundbaker Leah Penniman's (page 198) book *Farming While Black*, Groundbaker Dr. Liz Carlisle's (page 90) book *Healing Grounds*, and Groundbaker Dr. Gail Myers' (page 134) documentary film, *Rhythms of the Land*.

Discrimination in farming: How one million Black farmers lost their land: Mackenzie collaborated with Groundbaker Leah Penniman on a policy memo titled *Land Access for Beginning and Disadvantaged Farmers*. Here are some key facts and statistics from the report:

"In 1910, one in seven farmers were African American and held titles to approximately 16–19 million acres of farmland. Over the next century, 98% of Black farmers were dispossessed through discriminatory practices at the USDA and various federal farm programs. These farmers were often denied loans and credit, and lacked access to legal defense against fraud, resulting in a 90% loss of Black-owned farmland in the U.S. Today, 98% of private rural land is owned by white people, while less than 1% is Black-owned.

The USDA's systemic bias against Black and minority farmers is well documented and affirmed by the 2010 Pigford vs. Glickman class action lawsuit, which resulted in a $1.25 billion settlement. Black farmers continue to experience discrimination in access to credit, seeds, and other assistance, and they face foreclosure at six times the rate of their white counterparts."

It's also important to note that the settlements stemming from Pigford cover only specific recent claims of discrimination, and none stretching back to the period of the civil rights era, when the great bulk of Black-owned

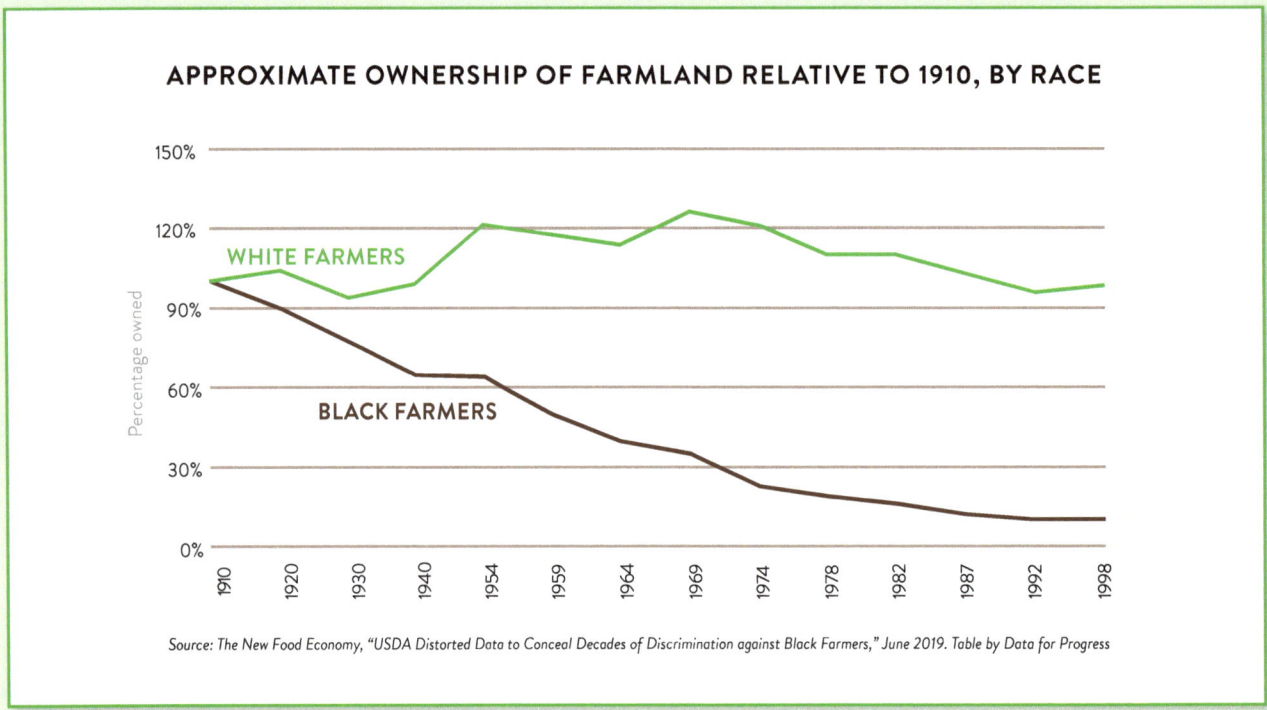

APPROXIMATE OWNERSHIP OF FARMLAND RELATIVE TO 1910, BY RACE

WHITE FARMERS

BLACK FARMERS

Percentage owned

150% 120% 90% 60% 30% 0%

1910 1920 1930 1940 1954 1959 1964 1969 1974 1978 1982 1987 1992 1998

Source: The New Food Economy, "USDA Distorted Data to Conceal Decades of Discrimination against Black Farmers," June 2019. Table by Data for Progress

farms disappeared. At the time of World War I there were 1 million Black farmers, and in 1992 there were 18,000.

In addition to the issues of land injustice and discrimination, as prices of farmland soar, large corporations are purchasing farms. For example, TIAA (Teachers Insurance and Annuity Association of America) holds more than 130,000 acres along the Mississippi River; over the course of a few years, TIAA has come to own almost the same amount of land as the African Americans who have been there for centuries, and they are only one big corporate landlord in the region. The accumulation of farms by TIAA and other corporations is yet another layer of history involving dispossession and land loss after the land was forcibly taken from Native Americans. Later, it was the labor of enslaved peoples from Africa who made the acres useful for intensive agriculture. Although many Black farmers were able to own land after Emancipation, a majority of these farmers were eventually stripped of their land.

Corporations, not farmers, have taken over the industry, as 3.2% of U.S. farms account for more than half the total value of the nation's agricultural production. Under this management, monocultural farming and unsustainable management practices such as persistent pesticide use are the norm, putting our food system and overall society in serious jeopardy.

In a 2019 article, *The Atlantic* cites investigations of the USDA where through its loan program, the department enacted illegal pressures that created huge wealth transfers from Black to white farmers, particularly in the period just after the 1950s. "In 1965," the article states, "the United States Commission on Civil Rights uncovered blatant and dramatic racial differences in the level of federal investment in farmers. The commission found that in a sample of counties across the South, the FmHA (Farmers Home Administration, a previous government agency) provided much larger loans for small and medium-size white-owned farms, relative to net worth, than it did for similarly sized Black-owned farms." The article goes on to state that Black land loss even included outright acts of violence or intimidation."

Today, 98% of private rural land is owned by white people, while less than 1% is Black-owned.

Other studies provide additional evidence about the unfair treatment of Black farmers. In a 1990 report, the House Committee on Government Operations concluded that the USDA "categorically and systematically denied minority farmers access and full participation in the multitude of Federal Government programs designed to assist them" and therefore is "directly responsible for the loss of land and resources these farmers have experienced."

Even the USDA called out its own wrongdoing. A 1996 USDA-commissioned study found that "97% of disaster payments went to white farmers, while less than 1% went to Black farmers," and that white men were given thousands more in loan packages than Black men. The agency's Civil Rights Action Team (CRAT) in 1997 determined that the USDA "took three times as long" to process Black farmers' loans as those of white farmers, and even when a loan was approved, it often "never arrives ... making it impossible for the farmer to earn any money from the farm." Calling out wrongdoing did not end the problem, as the USDA even lied in the 2014 Census of Agriculture, falsely sighting a 9% increase in Black farming to hide the discriminatory practices they were conducting.

so what can we do?

In a more just world, our food system would be based on a racially equitable, inclusive farm economy. Farmland would be seen not as a commodity, but as a tool that connects us back to the Earth and creates opportunity and stewardship among people from many backgrounds. There are organizations coming up with creative solutions to address the issue of land access. Here are examples of organizations that work at the local and national level:

- National Young Farmers Coalition
- California Farmer Justice Collaborative
- Soul Fire Farm
- Sustainable Iowa Land Trust
- Peninsula Open Space Trust
- California FarmLink
- National Family Farm Coalition
- Agrarian Trust
- Cooperation Jackson
- National Black Food and Justice Alliance
- Family Farm Defenders

Supporting organizations like these and advocating for legislation they support are great action steps. To find out more about this, check out our How to Get Involved section on page 230.

eggless salad sandwich

pre-prep: Soak cashews for 4 hours or quick soak (page 12)

prep time: 10 minutes • *makes*: 4–6 sandwiches

The secret ingredient in this Eggless Salad Sandwich is kala namak, otherwise known as Himalayan black salt. The high sulfur content in the salt gives off a savory umami taste that provides a very similar flavor to eggs. Your taste buds will be convinced you are eating the real thing!

ingredients

16 ounces firm tofu, drained

1 cup cashews, soaked and drained

1 tablespoon vegan mayo

1 teaspoon nutritional yeast

1 teaspoon lemon juice

2 teaspoons Dijon mustard

1 teaspoon distilled white vinegar

1 teaspoon Himalayan black salt (*kala namak*)

¼ cup water

¼ teaspoon black pepper

¼ teaspoon turmeric

1 teaspoon chopped sweet pickles (*or relish*)

directions

1. Drain and cut tofu into tiny cubes and set aside in a medium bowl.

2. Add all remaining ingredients except pickles to a small food processor and blend until creamy. Remove from blender and place in a separate medium bowl. Stir in pickles.

3. Add the cashew mix to the tofu bowl and stir to mix.

4. Spread on bread or crackers, and store remaining mixture in refrigerator.

dad's hearty chili

pre-prep: *If using dried beans, soak 6 hours or overnight and cook for 1 hour (page 11)*

prep time: *20 minutes* • **cook time**: *45 minutes*

My dad is so happy that one of his recipes made the book! On a cold night, it's the best meal to have with your family. Packed with protein, fiber, vitamins, and minerals from the variety of beans, this recipe is not only healthy and hearty, but so easy to create. The combination of flavors from the cumin, oregano, and chili powder also makes this chili extremely flavorful. Enjoy it with our Cornbread (page 132) or Sweet Potato Biscuits (page 150). *mackenzie*

ingredients

1 tablespoon olive oil

1 medium onion, diced

2 bay leaves

1 teaspoon ground cumin

2 tablespoons dried oregano

2 celery stalks, diced

1 red bell pepper, chopped

1 carrot, diced

1 jalapeno pepper, chopped

4 cloves garlic, crushed

1 can (*4 ounces*) chopped green chile peppers, drained

2 bottles (*18 ounces each*) crushed tomatoes

2 tablespoons chili powder
(*add more according to taste*)

½ tablespoon salt (*+ additional for seasoning*)

½ tablespoon pepper

1 cup kidney beans, cooked and drained

1 cup chickpeas, cooked and drained

1 cup black beans, cooked and drained

1 cup corn, frozen

optional additions

¼ cup quinoa, uncooked with ½ cup vegetable broth

meatless crumbles (*add ¼ cup of vegetable broth per cup of crumbles*)

serving ideas

avocado slices

Cauli/Cashew Sour Cream (*page 186*)

rice of choice

Cornbread (*page 132*)

Sweet Potato Biscuits (*page 150*)

directions

1. Heat olive oil in a large pot over medium heat. Stir in onion and season with bay leaves, cumin, and oregano.
2. Cook and stir until onion is tender, then mix in the celery, bell pepper, carrots, jalapeno peppers, garlic, and green chile peppers. Reduce heat to low and cook 5 minutes.
3. Mix the tomatoes into the pot. Stir in chili powder, salt, and pepper. If adding quinoa, add to pot with vegetable broth. Stir in the kidney beans, chickpeas, and black beans and bring to a boil. Reduce heat and simmer for 40 minutes.
4. Stir in the corn and continue cooking 5 minutes before serving. Serve with additional suggestions.

michel NISCHAN

Michel Nischan is a four-time James Beard Award–winning chef and cookbook author with over three decades of leadership experience advocating for a healthier, more equitable, and sustainable food system.

Nischan is the co-founder and chairman of Wholesome Wave, which is dedicated to nourishing neighborhoods by supporting increased access and affordability to healthy, fresh, and affordable fruits and vegetables for community members struggling with poverty. Following Paul Newman's lead, Nischan created Wholesome Crave, a food company selling plant-based food products to large-scale institutions to support the work of Wholesome Wave. In addition, he's a co-founder of Chef Action Network, a nonprofit committed to inspiring a network of influential chefs focused on food sustainability. He is founder and partner, along with the late Paul Newman, of the former Dressing Room restaurant. Wholesome Wave was instrumental in securing $250 million for the Gus Schumacher Nutrition Incentive Program in the 2018 Federal Farm Bill, expanding affordable access to locally grown fruits and vegetables. The James Beard Foundation honored Nischan as the 2015 Humanitarian of the Year.

> "I believe *food, as a single subject,* has more impact on human health, environmental health, economic health, and societal health than any other single issue."

Reflecting on his childhood, Nischan says it was his mother's cooking that inspired him to become a chef, and a fantastic chef at that. Nischan grew up eating fruits and vegetables straight from his mother's garden, and he also worked on his grandfather's farm every summer as a child. Now he realizes how much he took for granted his easy access to ripe, locally grown fruits and vegetables.

It wasn't until his son, Chris, was diagnosed with Type 1 diabetes that Nischan made the connection between food and human health. "As I changed the way we cooked at home so that Chris could have a long and healthy life," he says, "I realized that I wasn't reconciling with how I was feeding my customers. So, I changed the way I cooked in my restaurants, which then introduced me to a group of folks that were in the public health space who found me an anomaly. Wow, a three-star chef doing healthy food, they'd never heard of such a thing."

As his journey continued, Nischan started to discover the major systemic problems occurring throughout the food system involving lack of affordability for healthy food and equitable access to land for small-scale farmers. He also became aware of the prevalence of Type 2 diabetes, how many people were struggling with the condition, and the role income level played in that struggle. His dream is for people of all income levels to have the ability to "use food as an effective tool to prevent or avoid diet-related disease."

Nischan went on to found Wholesome Wave, which empowers underserved consumers to make better food choices by increasing affordable access to healthy produce. It works in two ways: First, by doubling the value of food stamps when spent on fruits and vegetables, and second, by working with doctors to prescribe produce.

Nischan adds that everybody has a right to really good quality, delicious basic ingredients. "A ripe tomato, a juicy cucumber, an apple, some fresh basil or dill or oregano. Everybody should be able to buy those basic ingredients for their family regardless of income."

Hey mom.

It's been so long. So damned long since you and I cooked together in your kitchen. We were always the last to sit at the table together to share a meal, after everyone else shoveled down your amazing country cooking. Far too long since I heard your gentle admonishment of me for putting too much flour in the dumpling dough, or heard you tell me you loved me while hugging me and smelling like fried chicken. Thank you for teaching me how to love unconditionally through food.

michel

Nischan also feels strongly about the unfortunate lack of financial incentives toward growing healthy foods. "I just think it's nuts that all of the crops that are not in our best health interests are the products that get all the financial support, the tax credits, and the subsidies. The things that are best for us get barely a penny."

In 2021, Wholesome Wave CEO Benjamin Perkins added the "FED" principle to their Produce Prescription Program, which stands for Fidelity, Equity, and Dignity. The FED framework allows Wholesome Wave to deepen, expand, and elevate a critically needed layer of racial and cultural sensitivity to future frameworks in the food as medicine movement. Wholesome Wave received a grant from Walmart, which will help make the FED a new national standard in addressing the racial and cultural inequalities faced by those who are food- and nutrition-insecure.

Nischan continues to fight for people to gain access to affordable locally grown food.

Wholesome Wave has expanded to 48 states in order to help make this happen: in Nischan's words, "Food is our common ground."

MICHEL NISCHAN'S
risotto-style summer heirloom farro

prep time: 15–20 minutes • **cook time:** 50 minutes • **makes:** 8 servings

Thankfully, the old-world grain farro made it to the new world so that we could enjoy this amazing dish. This dish resembles risotto, but, in deference to our American heritage, perhaps it should be called porridge. Semantics aside, it's a great way to use summer squash.

ingredients

4 tablespoons olive oil
1 large yellow onion, diced
8 fresh zucchini blossoms (*optional*)
1 cup uncooked farro
4 cups vegetarian broth
1 large zucchini, diced
1½ cups fresh sweet corn kernels

1 pint cherry tomatoes, halved
2 tablespoons unsalted butter
kosher salt and freshly ground black pepper
1 lemon, sliced

To make it plant-based:
• vegan butter can be substituted for butter

directions

1. Heat 2 tablespoons of the oil in a large, deep skillet over medium-high heat. Add the onion. Cook for 4–5 minutes, or until softened, but not browned. Add the zucchini blossoms, if using, reduce the heat to medium, and cook for about 1 minute.
2. Remove the blossoms to a plate and keep warm. Add the farro. Cook for 2–3 minutes just to coat with the oil and mix with the onions.
3. Pour the stock into the skillet. Stir the farro and onions. Bring to a boil over medium-high heat. Reduce the heat to medium.
4. Simmer, stirring frequently, for 40–45 minutes, by which time the farro should be tender and the stock evaporated. Add more stock if needed during cooking to keep the farro moist.
5. Heat the remaining 2 tablespoons of oil in another large skillet over medium-high heat. When hot, add the zucchini and corn kernels. Cook for 10–12 minutes, or until the vegetables brown.
6. Add the tomatoes. Cook for about 1 minute to warm through. Add the farro and toss to mix. Add the butter and stir until melted. Season to taste with salt and pepper.
7. Spoon the farro onto each of 8 plates. Garnish each plate with a zucchini blossom.

On the end of every growing zucchini, there is a yellow-orange flower blossom, which is its own separate vegetable!

mac's favorite
mac and cheese

pre-prep: 50 minutes to bake potatoes. Soak cashews for 4 hours or overnight, or quick-soak (page 12)

prep time: 20 minutes • *cook time*: 10 minutes • *serves*: 5–6

*O*h, the classic...every kids' favorite meal. We couldn't have a comfort-food cookbook without mac and cheese! Although that boxed Kraft tasted so good when I was a kid, it lacked nutritional value and contained lots of fillers, as well as artificial flavors and colors. We've revamped this all-time favorite to mimic the taste and texture of the boxed version, but instead use nutritional ingredients such as sweet potatoes, coconut milk, carrots, and cashews. No longer loved by only kids, this plant-based powerhouse is destined to be a family favorite! *mackenzie*

ingredients

1 box pasta, 8 ounces

2 Yukon Gold potatoes

½ sweet potato

3 small carrots

½ onion

1 cup cashews, soaked and drained

¾ cup coconut milk

juice of 1 lemon

1 teaspoon paprika

1 teaspoon smoked paprika

2 teaspoons salt

5 tablespoons nutritional yeast

½ cup water from cooked pasta

topping suggestion

Coconut Bacon *(page 171)*

directions

1. Cook pasta as directed. Drain, and save ½ cup water. Set pasta aside.
2. Preheat oven to 375°F. Pierce potatoes, wrap in foil and bake for 45–50 minutes until tender.
3. In a medium pot, steam carrots and onion in a steamer basket on medium heat for 6–8 minutes, until tender (see page 12 for steamer directions).
4. Add all ingredients except pasta water to blender and blend on medium high for 45–50 seconds.
5. Pour blended ingredients into a medium pot and simmer on low for 10 minutes. If sauce is too thick, add reserved water from cooked pasta.
6. Pour desired amount of sauce over pasta. Serve with Coconut Bacon (page 171) and grilled veggies of your choice.
7. Store remaining sauce in refrigerator.

aileen SUZARA

Aileen Suzara is a Filipino-American chef, educator, writer, mama, and activist who combines her love of food with experiences shaped by sustainable farming, environmental justice, and community health advocacy.

Her interest sparked by the rise of chronic disease among Filipino Americans and communities of color, Suzara pursued a master's degree in public health nutrition at UC Berkeley, with a focus on reclaiming her community's foodways for wellness. She launched a food project titled Sariwa (*Fresh*) that reimagines plant-based Filipino foods and incorporates the healing strengths of her community's heritage. Sariwa was a member of La Cocina's acclaimed incubator and Suzara has offered workshops and pop-up community meal projects to audiences across California and Hawai'i. She has engaged students of all ages at schools, universities, organizations, and hospitals. Notable collaborations include Sama Sama Cooperative, Kaiser Permanente, and the Rooted Recipes Project. Suzara was recognized as a Bon Appetit Healthyish Honoree and a Castanea Fellow.

Dear families of Sama Sama Cooperative,

I am grateful for our summers cooking together and feeding one another. For your curiosity and sense of play when in the kitchen. For your tabi tabi po and respect for plants and for the unseen world when in the garden. For your fierce sense of justice for people and Mother Earth in how you move through the world. Thank you especially to our young ones, for teaching us new ways to remember and rediscover the healing within our roots. You are the dream of our ancestors!

Yours in food,

tita aileen

The child of immigrants, Suzara grew up in the 1980s facing racist stereotypes and societal pressures to assimilate. Her parents were advised to shed their family's foods and even languages in order to gain acceptance. Adapting to "American" ways meant that processed and canned goods like Spam were staples in their pantry. Suzara explains that her family was trying "to embrace this idea of Americanized convenience food. I only realized later that these types of foods were not ones my grandparents would have recognized."

At eight years old, Suzara desired to reconnect to her Filipino heritage. One day, she discovered a Filipino cookbook on her family's bookshelf. As she flipped through the pages, she opened a doorway to her future. Her curiosity took over and she wanted to learn more about the book's recipes, her culture, and her ancestral history. She asked herself, **"What are these ingredients and what are their significance to my family?"** Thankfully, Suzara's parents never fully shed their connection to their Filipino ancestry, and as Suzara reconnected to her roots, they did too.

Suzara conversed with her family about their memories, where they came from, and what sustained them. Growing up in Hawai'i also illuminated Suzara's curiosity about health, wellness, and sustainable food systems. While her family expected her to work in the healthcare field, Suzara found that cooking and farming were also powerful ways to be engaged in health. With this clarity, she engaged with her lifelong passion.

Aileen offers these words of wisdom around food and diet:

- **"First, make sure we re-center the culture in agriculture. That's one way that we can have more diverse and inclusive food systems."**

- **"Next, be creative when it comes to food, and have an element of delight and beauty and fun."**

- **"Finally, encouragement from friends, support from family, and connection with our broader community can support long-term changes in our food relationships."**

Suzara focuses on harnessing the power of Filipino food. Given the stress and diet-related health disparities within her own community, she feels it is her responsibility to make an impact. Suzara reminds us that diet-related diseases are not problems rooted in individual choices, but are symptoms of systemic inequities. She firmly believes that food is medicine and that health is about the community as a shared collective, rather than individuals. Suzara says her work encourages, **"conversation about a community that's very much here, but isn't always visible in the mainstream media narrative."**

Suzara finds her work rewarding when the food she makes encourages someone to uncover a food memory, have deeper conversations with their loved ones, or reflect on their own cultural roots and personal food journeys. These responses launch individuals towards empowerment and political engagement. "Looking to our roots can offer perspective on the ways our ancestors survived in the past as we navigate this current moment ... it's actually one of the most innovative things that we can do right now."

"Looking to our roots can offer perspective on the ways our ancestors survived."

AILEEN SUZARA'S

ginataang sitaw at kalabasa
(long bean and squash in coconut milk)

prep time: 15 minutes • *cook time:* 25 minutes • *makes:* 3–4 servings

ingredients

1 tablespoon coconut or olive oil

1 small onion, diced

4 cloves garlic, minced

1 tablespoon ginger, grated

3½ cups coconut milk

3 cups kabocha squash, cut into 1½–2" cubes
  (The green skins are edible!)

1½ cups long beans, cut into 2" segments

salt and black pepper

1 lemon or calamansi

1 tablespoon soy sauce or amino acids, to taste

optional

1 bird's eye or green chile, sliced scallions,
  crispy garlic chips

variations

Substitute spinach, fresh moringa, or seasonal
  favorites like okra for long beans; add baked tofu
  or other proteins for a more substantial entrée.

directions

1. Heat oil in a heavy-bottomed pot or deep saucepan over medium heat. Add onions and cook until soft, about 5 minutes. Add a sprinkle of salt.

2. Add garlic and ginger. Sauté until fragrant.

3. Add coconut milk and squash, cover, and simmer until squash is tender, about 12 minutes.

4. Add long beans and chilies. Cover and simmer just until long beans are tender, but still crisp, about 5 minutes.

5. Season to taste with salt, pepper, and a squeeze of citrus. Enjoy with a side of delicious garlicky fried rice or steamed rice.

6. As with most traditional Filipino dishes, season to taste. Deepen the savory flavor with soy sauce or amino acids to taste and any preferred toppings like crispy garlic. Traditionally, bagoong (shrimp paste) is added for a briny punch—this can be replicated with vegan shrimp paste (yes, this exists!).

matt's loco moco

pre-prep: 20 minutes to make a Vegan Fried Egg (page 177),
20 minutes to make Cashew Mushroom Gravy (page 185), plus overnight soak for cashews (page 12)
prep time (for mushroom steak): 25 minutes
cook time: 20 minutes • *makes*: 7–8 servings

My brother, Matt, has always loved the popular Hawai'i favorite "Loco Moco," which is a hamburger patty on top of white rice, topped with an egg and gravy. Since Matt became a vegan, he hasn't been able to replicate his all-time favorite until now. We worked together to create an awesome Mushroom Steak, combining mushrooms and tofu, and adding some tasty sauces and spices until we found the perfect flavor and texture. With the help of black salt (kala namak), we were also able to re-create a tasty Vegan Fried Egg (page 177) to complete this incredible meal. Put it all together, and you have a Loco Moco that everyone can enjoy! *mackenzie*

ingredients for mushroom steak

½ cup bread crumbs

¼ cup + 1 tablespoon gluten-free flour

2 tablespoons sweet rice flour

¼ teaspoon pepper

1½ tablespoons cornstarch

16 ounces firm tofu, drained

1 teaspoon olive oil

2 cups mushrooms, king oyster or
   portobello, finely diced

1 small onion, finely diced

3 cloves garlic, minced

1 flax egg (*page 10*)

3 tablespoons vegan oyster sauce
   (or 1 tablespoon worcestershire sauce, 1 tablespoon
   hoisin sauce, and 1 tablespoon coconut aminos)

1 tablespoon tamari

vegan butter or olive oil for frying

serve with

white basmati rice • Vegan Fried Egg (*page 177*) • Cashew Mushroom Gravy (*page 185*)

Recipe continued on page 82

directions

1. In a large bowl, add bread crumbs, gluten-free flour, sweet rice flour, pepper, and cornstarch. Mix well and set aside.

2. Cut tofu into 2" cubes. Place cubes in cheesecloth or nut bag and squeeze out as much moisture as possible. Tofu will be very crumbly. Add tofu crumbles to a medium bowl.

3. Add 1 teaspoon of olive oil to a large frying pan. Over medium heat, cook mushrooms for 3–4 minutes, until mushrooms release water. Add onions, cook for 3 more minutes. Add garlic and cook for 1 more minute.

4. Add mushroom mix to medium bowl with tofu crumbles. Add flax egg, oyster sauce, and tamari. Gently mix together.

5. Add mushroom/tofu mixture to flour mixture in the large bowl and stir well.

6. Make 7–8 patties and place on a baking sheet. In a large skillet, heat butter or olive oil and fry 3 patties at a time for 3–5 minutes on each side until golden brown.

7. To assemble, add a scoop of white basmati rice to a serving bowl. Top with mushroom steak and a Vegan Fried Egg (page 177). Add a heaping serving spoon full of Cashew Mushroom Gravy (page 185) and voilà! You've created Matt's favorite meal (also a favorite with most of the people of Hawai'i).

miso meatballs

pre-prep: If using dried chickpeas, soak 6 hours (or overnight) and cook for 1 hour (page 11).
prep time: 15 minutes • *bake time*: 25 minutes • *makes*: 15–18 meatballs

Meatballs are enjoyed around the world by many different cultures. When developing this recipe, we had fun experimenting with different regional versions, each with its own array of flavors and spices. This variation is our favorite, and it tastes great with our Pesto Sauce (page 89), Cashew Mushroom Gravy (page 185), Alfredo Sauce (page 180), or Marinara Sauce (page 178). Enjoy in a wrap, over pasta, or with rice!

ingredients

½ cup walnuts

½ cup oats

2 tablespoons nutritional yeast

¼ teaspoon salt

7 ounces firm tofu, drained

1 tablespoon miso paste

½ cup chickpeas

1 flax egg *(page 10)*

1 teaspoon vegan oyster sauce

1 teaspoon tamari

2 teaspoons sesame oil

1 medium onion, diced

2 garlic cloves, minced

directions

1. Preheat oven to 350°F. Line a baking tray with parchment paper and set aside. Place walnuts, oats, nutritional yeast, and salt in food processor and pulse until crumbly. Transfer ingredients to a medium bowl. Place tofu, miso, chickpeas, flax egg, oyster sauce, and tamari in the food processor and blend for about 10–15 seconds. Add ingredients to medium bowl.

2. Heat 1 teaspoon sesame oil in small pan, fry onion for 3–4 minutes, then add garlic. Stir until cooked.

3. Add onion mixture to the medium bowl and mix together. Mixture should be sticky.

4. Make 15–18 meatballs out of mixture and place on baking sheet. Cook meatballs for 15 minutes, then turn and bake for another 10 minutes. Heat 1 teaspoon sesame oil in frying pan over medium/high heat and fry meatballs for 3–4 minutes before serving. Leftovers can be stored in the fridge.

* For photo of the **meatballs**, see page 88.

pepperoni pizza pinwheels

pre-prep: If making Mozzarella Cheese: 20 minutes (page 184)

If making Parmesan Cheese: 1 hour and 10 minutes (page 183)

If making Marinara Sauce: 40 minutes (page 178), Eggplant Pepperoni: 40 minutes (page 176)

prep time: 1 hour 15 minutes • **cook time**: 20 minutes • **makes**: 20–22 pinwheels • **serves**: 5–6

You can choose the amount of time you want to spend on this recipe. If you want to make EVERYTHING from scratch (and we mean everything, including the sauce and cheeses), we suggest you take a look at this recipe at least a day or two before you plan to make it so you can prepare adequately. The dough can be made earlier in the day and then refrigerated after it rises and is rolled out so it's fresh out of the oven when you are ready to eat. If you prefer to spend less time on this, a lot of the ingredients we call for can be purchased at your local market. The pinwheels are so versatile…they make for a perfect potluck dish, afternoon snack, or fun family dinner. Enjoy!

ingredients

For the dough

1 cup water, 110–115°F

1 tablespoon honey

2¼ teaspoons active dry yeast

2 ½ cups gluten-free flour

1 teaspoon salt

1 tablespoon coconut sugar

1 tablespoon olive oil + extra oil to grease bowl

extra flour for dusting

For the pinwheels

1 tablespoon olive oil

4 cloves garlic, minced

2 cups fresh spinach, cut

salt and pepper to taste

Eggplant Pepperoni (page 176)

1 cup plant-based Mozzarella Cheese (page 184)

½ cup plant-based Parmesan Cheese (page 183)

1 cup Marinara Sauce (page 178)

extra olive oil to grease baking sheets and
  to brush onto pinwheels after baking

Recipe continued on page 86

directions

pizza dough

1. Warm water to 110–115°F. Add honey and yeast, stir. Let mixture sit for 8–10 minutes until yeast proofs.
2. In a large bowl, add flour, salt, and sugar. Whisk to combine. Add olive oil and yeast mixture and mix with a large spoon until well-combined.
3. Use your hands to gently knead mixture. Place dough in a greased bowl. Roll dough around to lightly cover it with oil. Cover bowl with a clean towel and set aside to rise for 1 hour.
4. Cover each baking sheet with plastic wrap and dust with flour.
5. Divide dough in half. Place each dough half on its own baking sheet. Using a rolling pin, flatten dough on sheet into a rectangle shape. Do this for both halves. The shapes do not have to be perfect rectangles. If you are ready to make the pinwheels then let's get started! Otherwise, if you prefer to wait until you are about an hour away from serving, leave the dough halves on the baking sheets, cover with extra plastic wrap and place in refrigerator. Take dough out about ½ hour before you are ready to roll your pinwheels.

pinwheels

1. Preheat oven to 425°F. In a large frying pan, heat olive oil over medium heat. Add garlic, sauté for a few minutes. Add spinach and toss with tongs for two minutes, until wilted. Season with salt and pepper.
2. Evenly spread spinach mixture onto each dough rectangle. Add eggplant pepperoni and drops of mozzarella cheese. Season with parmesan cheese.
3. Starting at the long end, tightly roll dough up. After dough is completely rolled, gently slice each roll into 10–12 even rounds.
4. Remove plastic wrap with sliced rolls to the side and grease the baking sheets with oil. Lay pinwheels down evenly on the sheet. Set baking sheets aside for another 15 minutes in a warm, draft-free area to let the dough rise a bit more.
5. Bake pinwheels for 15–20 minutes, until tops and edges start getting golden brown.
6. Remove from oven and brush with olive oil. Let pinwheels cool. Serve with Marinara Sauce (page 178)

why not wheat?

People have harvested and eaten wheat for over 5,000 years and it hasn't become a health problem until recently. So, what has changed? Back in the 1960s, a plant scientist named Norman Bourlag, known as the "Father" of the Green Revolution and recipient of the Nobel Peace Prize, developed a strain of wheat called dwarf wheat that could produce higher yields. This has become what we know today as wheat, as almost all commercial flour since the 1960s is derived from this developed hybrid. Modern wheat is much higher in gliadin, which is linked to celiac disease. Modern wheat also depends on synthetic herbicides and fertilizers, and it is treated with bleach, and chemical agents that are banned in Europe, including potassium bromate, which is linked to cancer, and azodicarbonamide, which is linked to asthma, skin irritation, and allergies. Even if you don't eat wheat, a lot of this hybridized grain is fed to animals and trickles down to those who eat animals. To avoid the health issues caused by today's wheat, we chose to focus on wheat alternatives that are delicious, nutritious, and hopefully less likely to cause health issues.

pesto pasta

pre-prep: If making Miso Meatballs, 40 minutes (page 83)
prep time: 30 minutes • *cook time*: 10 minutes • *serves*: 4

Making your own pesto is easy and tastes incredible with the fresh basil leaves and lime juice. The standard pesto recipe is not dairy-free because it usually includes parmesan cheese, but in this recipe you can use our Cashew Parmesan Cheese (page 183) to make vegan pesto! Add Miso Meatballs to the recipe for protein (page 83).

ingredients

pesto sauce

1 tablespoon olive oil for sautéing garlic and nuts
3 cloves garlic, minced
⅓ cup pine nuts
⅓ cup walnuts or macadamia nuts
2½ cups basil, stemmed and washed
2 tablespoons Cashew Parmesan Cheese (*page 183*)
1 tablespoon lime juice
⅓ cup olive oil
½ teaspoon salt
½ teaspoon pepper
1–2 tablespoons water, depending on consistency

pasta

8–12 ounces pasta of choice
1 tablespoon coconut oil
1 onion, sliced
1 cup mushrooms, sliced
2 cups grape or cherry tomatoes, sliced in half
salt and pepper to taste
Cashew Parmesan Cheese (*page 183*)
Miso Meatballs (*page 83*)

directions

1. Add 1 tablespoon olive oil in medium pan. Add garlic, pine nuts, walnuts or macadamia nuts to pan and sauté for 3–5 minutes. Set aside and let cool.

2. Once cooled, add to a food processor or blender with all other pesto ingredients except the water and blend until completely smooth. If the sauce is too thick, add 1–2 tablespoons water. Set sauce aside.

3. Cook pasta according to directions. Drain pasta and set aside in a large bowl. Add ½ cup pesto sauce to pasta and toss lightly to coat. Set aside remaining pesto sauce.

4. Heat oil in a large frying pan over medium-high heat. Add onions and mushrooms, and sauté for 2–3 minutes. Add tomatoes, salt, and pepper and sauté for another 3 minutes until the tomatoes are cooked.

5. Add cooked pesto pasta to the large pan and toss together to mix. Add cooked Miso Meatballs and toss. Add remaining pesto sauce if necessary. Serve directly from pan.

DR. *liz* CARLISLE

Assistant Professor Dr. Liz Carlisle teaches courses on food and farming in the Environmental Studies Program at UC Santa Barbara.

Born and raised in Montana, Carlisle got hooked on agriculture while working as an aide to organic farmer and U.S. Senator Jon Tester, which led to a decade of research and writing collaborations with farmers in her home state. Carlisle has written three books about re-generative farming and agroecology: *Lentil Underground*, which won the 2016 Montana Book Award, *Grain by Grain* (with co-author Bob Quinn), and most recently, *Healing Grounds: Climate, Justice, and the Deep Roots of Regenerative Farming*. She is also a frequent contributor to academic journals and popular media outlets, focusing on food and farm policy, incentivizing soil-health practices, and supporting new-entry farmers. Her work has been featured in the *New York Times, Business Insider,* and the *Los Angeles Times*. Carlisle holds a Ph.D. in Geography from UC Berkeley and a B.A. in Folklore and Mythology from Harvard University. Prior to her career as a writer and academic, she spent several years touring rural America as a country singer.

Carlisle was originally inspired by her grandmother, who had powerful insights into humans' connection with the natural world. As a young girl, her grandmother lost her farm due to her father's use of damaging plowing techniques and the nationwide Dust Bowl phenomenon. Understanding her grandmother's beliefs and personal stories was really what "planted the seeds" for Carlisle's journey to find a more sustainable way of farming.

By traveling throughout America as a country singer, Carlisle learned how different communities stewarded their land and created a livelihood. She realized that it wasn't simply a question of choice that governed how people managed their land. Farmers were also constrained by the Farm Bill and the economics of the food system.

She decided that foundational policy issues needed to be changed before anything else.

Carlisle stopped traveling and applied for a job with fellow Montanan Jon Tester, a U.S. Senator and organic farmer with policy proposals that addressed issues within the food system. Tester used his platform to ignite positive change, so when Carlisle got the opportunity to work with him, she took it. In this role, she highlighted the value of organic farming while prioritizing the right of rural communities and family farms to stay on their land. As a legislative correspondent for agricultural and natural resources, Carlisle was assigned to stay in touch with all of the local farmers. Her love for sharing stories, therefore, stayed present in her life. During her time at the Senate office, Carlisle noted that Tester had sparked a movement of farmers who assembled to decide which crops were necessary to add to farms in order to stop relying on chemicals. Later, this movement would help her learn the value of lentils and inspire her first book.

Carlisle decided to pursue her interest in this movement and entered grad school at UC Berkeley, where she wrote her thesis on sustainable agriculture and organic farming. She also connected with David Oien, a third-generation

Montana farmer and co-founder of Timeless Foods. He was teaching people how to grow crops, such as lentils, without the need for chemicals.

After conversing with Oien, Carlisle began writing *Lentil Underground*, which focuses on the value of leguminous crops like lentils and how they contribute to nitrogen fixation, a process that provides biological fertilizer for plants. She knew that this book would be helpful for the

Dear Grandma Helen,

A resilient daughter of the Dust Bowl, you taught me to listen to the land. If you want wisdom, girl, get your hands dirty. If you want justice, go deep. Don't be afraid to jump some fences. A little bit wild is a good thing.

liz

movement towards sustainable agriculture because she had, "zeroed in on this origin point for something really important that not enough people knew about."

On the importance of plant-based foods, Carlisle states, "It's also about the wisdom embedded in a lot of the culinary traditions from our ancestors. Examined in this light, plant-based cuisine is not seen as an alternative food culture, but actually, a couple generations before industrial food was so aggressively marketed to us, we could find an abundance of plant-based food in grandma's kitchen or great-grandma's kitchen."

Carlisle observes that people have long been exposed to a **"strong narrative that we can't feed the world with ecological practices or that there are deficits in our ability to produce adequate food on a global basis.** And so this is the argument made for an agriculture that uses toxins," she says. "An agriculture that's more consolidated. An agriculture that uses proprietary technology ... And yet, if we're going to work on global food security, if we really are concerned about having enough to go around, [we know] that the big levers are food waste and more plant-based protein rather than animal-based protein. [This is] because you're losing calories up the food chain as you feed a plant to an animal. You know, [Frances] Lappé calculated that you're losing [about] 90% of the energy embedded in that plant in the process of putting it through an animal."

Carlisle is open minded in her beliefs about animal protein. "The history of vegetarianism and veganism," she says, "obviously feels really aggressive to a lot of ranchers and in some cases it has been, and probably unfairly. At the same time ... I also respect the fact that there are people who have strong spiritual beliefs that animal agriculture isn't appropriate or ethical, and there needs to be room for that in the conversation too. I think [it's about] a joyful food system ... I think humanity has a history of food as something to celebrate. It's a way of celebrating our connection to each other and the natural world. Like you think about what people say right before they eat. You know, those are some of the most beautiful, insightful things ... So for me, the work is really about centering that joy and celebration as we move towards these synergies between sustainability and social justice."

In addition to her writing, Carlisle works as a teacher for the younger generation, whom she congratulates for facing challenges like climate change and racial injustice head on. She believes that, "the promise of getting engaged in transforming the food system is that it involves imagining and then actually manifesting the kind of society we could be. It's a chance to do really positive work on that intersection of sustainability and social justice." She tells her students that, "the earth isn't one more problem for you to carry. It's like the ally that's going to help you through this ... Having that strong connection ... gives you something to ground in."

In *Healing Grounds*, her latest book, Carlisle turns her attention to Indigenous, Black, Latinx, and Asian-American farmers who are growing food by turning back to the methods used by their ancestors. Some are rejecting monoculture to grow corn, beans, and squash, as farmers in Mexico have done for centuries. These farmers are enriching the land by restoring native prairies, nurturing beneficial fungi, rotating crops, improving soil health, and repairing the natural carbon cycle.

As stated in the description by publisher Island Press, cultivating a system built on this holistic kind of regenerative agriculture, not merely a set of technical tricks for storing CO_2 in the ground, will inevitably require reckoning with our country's agricultural history and the harsh reality that many farmers of color have been prevented from owning land or building wealth. The task of creating systemic change to correct these policies presents a huge challenge, but also offers the hope of restoring farmlands, communities, and planetary health.

DR. LIZ CARLISLE'S
spicy lentils

prep time: 5 minutes • **cook time:** 35 minutes • **makes:** 2-4 servings

This recipe is adapted from a recipe by Fetlework Tefferi, founder of Brundo and Café Colucci. Most of these ingredients are optional. The only things you really need are an onion, the lentils (split peas or mung beans work too), and a spice blend. A little oil makes a big difference, but I've still enjoyed this dish without it.

ingredients

1 red onion, minced

¼ cup olive oil

1 tablespoon of your favorite spice blend
*My favorites are berbere and afrenje from Brundo;
a nice curry powder or chili powder works well too.*

1 tablespoon garlic, minced

Water
*Full teapot works best; cold water is a
bit slower, but works fine too.*

1 cup organic split red lentils
I use Petite Crimson lentils from Timeless Natural Food.

½ teaspoon black pepper

salt

directions

1. Sauté the minced onions and olive oil in a large pot until the onions are translucent, adding water as needed to keep the onions moist.
2. Add your spice blend and garlic and sauté for 2 minutes more.
3. Add enough water to submerge the onions a half inch.
4. Stir in the lentils and cook for 20–30 minutes, stirring frequently and adding water to keep the sauce moist. The liquid should cover the lentils completely.
5. When the lentils are soft, turn the heat down to low and add black pepper and salt to taste.
6. Serve with your favorite whole grain—rice, barley, quinoa, or whatever you have handy! If you happen to have an Ethiopian market nearby that sells injera (Ethiopian flatbread), that is a special treat!

* Lentils, which are members of the legume family, have been eaten by people since Neolithic times and were one of the first cultivated crops. Lentils are also great for the soil, as they are nitrogen fixers.

pulled "pork" sliders

pre-prep: If making Texas BBQ Sauce, 10 minutes (page 179),

If making Sweet Potato Buns, 2 hours, 20 minutes (page 146)

prep time: 30 minutes • cook time: 6-7 minutes • serves: 5-6

Why should giving up meat mean you have to skip out on fun meals? Combining our tangy Texas BBQ Sauce with this "meat" mixture, topped with creamy coleslaw all on a Sweet Potato Bun, we show how it's possible to keep the flavors we love in a way that's better for us and for the planet.

ingredients

coleslaw

2 cups shredded green cabbage

2 cups shredded red cabbage

¼ cup vegan mayonnaise

salt/pepper to taste

¼ cup apple cider vinegar

"meat" mixture

1 tablespoon olive oil

2 large portabello mushrooms, sliced in strips

1 eggplant, sliced in strips

1 teaspoon garlic powder

1 teaspoon onion powder

½ teaspoon pepper

1 teaspoon salt

1 teaspoon paprika

1 teaspoon chili powder

½ teaspoon garlic powder

¼ cup Texas BBQ Sauce + extra for topping (page 179)

buns

Sweet Potato Buns (page 146)

condiments

pickles, extra barbecue sauce

directions

1. Prepare coleslaw in a large bowl by combining cabbage, mayo, salt, pepper, and apple cider vinegar. Mix well and refrigerate.
2. Prepare the "meat" by heating oil in a large frying pan. Stir in mushroom and eggplant slices, and sauté for a minute. Add spices and continue stirring for about 6–7 minutes. Stir in ¼ cup barbecue sauce. Remove from heat.
3. Slice buns in half, and toast if desired. Add a scoop of "meat" to each bun. Add more barbecue sauce if desired. Top with coleslaw and pickles.

southwest salad

pre-prep: *If using dried black beans, soak for 6 hours or overnight and cook 1 hour (page 11)*
prep time: *20 minutes* • Serves: *4-6*

Most people think of salad as a side dish, but this recipe is a whole meal in itself, filled with flavor, protein, fiber, and nutrients. It's easy to assemble, and we encourage you to top it off with our amazing Ranch Dressing (page 188). You are really eating the rainbow with this one!

ingredients

1 head romaine lettuce, chopped

1 cup corn, canned or frozen and defrosted

1 red bell pepper, chopped

1 cup cherry or grape tomatoes, cut in half
   We like them cut lengthwise.

1 cup black beans, cooked or canned, drained

⅓ cup pumpkin seeds

½ cup grated plant-based cheese
   We used vegan pepper jack cheese.

1 avocado, cut in chunks

corn tortilla chips, broken up

cilantro and shredded carrots to garnish

extra avocado, cut in slices

topping ideas

Ranch Dressing (*page 188*) • Coconut Bacon (*page 171*)

directions

1. Add salad ingredients to a large bowl, except tortilla chip pieces, cilantro, avocado slices, and carrots. Toss to combine.

2. Top salad with tortilla chip pieces, avocado slices, cilantro, carrots, and dressing. Sprinkle with coconut bacon (optional).

* **Known as pepitas in Mexico**, pumpkin seeds are a perfect inclusion to a salad that adds protein, antioxidants, manganese, magnesium, iron, and zinc.

sweet potato chickpea bbq wraps

pre-prep: If using dried chickpeas, soak 6 hours or overnight and cook for 1 hour (page 11),

If making Texas BBQ Sauce, 10 minutes (page 179),

If making Caesar Dressing, 10 minutes (page 189)

prep time: 40 minutes • **cook time**: 40 minutes • **makes**: 4–6 wraps

These wraps are bursting with flavor! The combination of sweet potato and BBQ-covered chickpeas, drizzled with Caesar Dressing and paired with crunchy cabbage is truly incredible. To make this recipe as simple as possible, prepare some of the staples beforehand, such as the Cassava Flour Tortillas (page 149), Texas BBQ Sauce (page 179), and Caesar Dressing (page 189). Or, to make it even simpler, you could purchase these additions at your local market. This recipe is a great go-to for lunch or dinner.

ingredients

4 cups sweet potatoes, uncooked, chopped into cubes (½" pieces)

1 tablespoon olive oil

1 medium onion, chopped

1 bell pepper, any color, chopped

1½ cups chickpeas, cooked or canned, drained

¾ cup Texas BBQ Sauce *(page 179) + extra for spreading*

4–6 large tortillas of choice
(page 149 for Cassava Flour Tortillas)

1 cup corn, either canned or frozen, steamed for 5 minutes

2 tablespoons pumpkin seeds

2 cups shredded cabbage

Caesar Dressing to drizzle *(page 189)*

directions

1. Preheat oven to 415°F.
2. Line a baking sheet with parchment paper. Add sweet potato cubes and bake for 15 minutes. Then, flip potatoes and bake another 15 minutes, until golden brown.
3. While potatoes are roasting, add oil to pan and heat to medium high. Add onions and bell pepper and sauté for 5 minutes. Set aside.
4. Add chickpeas to a small pot with ¾ cup of barbecue sauce over medium heat. Cook for 5 minutes until hot.
5. Spread extra barbecue sauce on tortillas. Add chickpea/barbecue sauce mixture on top of sauce. Add sweet potatoes, corn, and onion/bell pepper mixture. Top with pumpkin seeds and shredded cabbage.
6. Drizzle Caesar Dressing (page 189) over each tortilla. Roll up each tortilla and serve. Store extra sauces in refrigerator for later.

josé ANDRÉS

José Andrés is a Spanish-American chef, restaurateur, best-selling author, educator, and founder of World Central Kitchen, a nonprofit committed to supplying hot meals in the aftermath of natural disasters.

Andrés owns over 30 restaurants linked to the company ThinkFoodGroup, which he hopes will, "change the world through the power of food." In honor of his culinary skills and humanitarian actions, Andrés has earned several awards and distinctions. In both 2012 and 2018, he was named one of *Time* magazine's 100 Most Influential People. The James Beard Foundation awarded him the title of "Outstanding Chef" in 2011 and also recognized him as "Humanitarian of the Year" in 2018.

José Andrés founded his nonprofit, World Central Kitchen, in 2010 after he and his wife, Patricia, envisioned an organization that could create solutions for hunger and poverty in times of crisis. Through the provision of culinary training programs, World Central Kitchen has helped strengthen and empower communities when they need it most.

In the wake of Covid-19, World Central Kitchen served fresh meals in 400 cities, providing 40 million meals across the U.S., Spain, Indonesia, and the Dominican Republic. They also disbursed $167 million to local restaurants serving people in their communities. The work accomplished by World Central Kitchen, Andrés notes, goes much deeper than the photos that the general public sees. "We are about adapting with what's on the ground," he says. In addition, Andrés supports small farmers, fishers, and local businesses to help develop stronger economies in rural areas.

Andrés is inspired by many people and parts of his life. One of his biggest sources of inspiration is a quote from Robert Egger, founder of DC Central Kitchen, who said, "giving should not be about the redemption of the giver, but about the liberation of the receiver."

For Andrés, this "really put things in perspective … He [Egger] lives to that standard."

In the wake of Covid-19, World Central Kitchen served fresh meals in 400 cities, providing 40 million meals across the U.S., Spain, Indonesia, and the Dominican Republic.

Dear Clara,

When I first arrived in Washington, DC to open my restaurant, Jaleo, I would look across the street to the old Missing Soldiers Office, where you cared for the wounded during the Civil War. You were just a humble nurse putting your skills at the service of others—but your example inspired a movement of ordinary citizens giving back in simple ways... an ocean of empathy. Without ever knowing it, you inspired me to dream that cooks like me could use our skills in times of crisis to feed not just the few, but the many.

José

José adds that many people are simply focused on the perception that you're doing good. "It is the feeling that you are redeeming yourself, but you are not liberating anybody." José wants others to focus on the act itself, not on the perception.

To give an example of Andrés' typical work schedule, in 2021, immediately after the tragic 7.2 magnitude earthquake that shattered Haiti, World Central Kitchen hit the ground to serve fresh meals to families affected in Les Cayes and Jeremie, two of the cities that were most impacted.

One month later, a volcano erupted in La Palma, one of Spain's Canary Islands, covering 380 acres of land with lava, forcing the evacuation of 6,000 people. World Central Kitchen immediately responded by providing meals and first responders, who were working around the clock.

In 2021, Jeff Bezos awarded Andrés a $100 million grant, which Andrés will use to continue the work of World Central Kitchen and reach his goal—to double food aid around the world. He has allocated some of those funds to set up mobile kitchens at the Ukraine-Poland border to provide meals for Ukrainians fleeing the Russian invasion and will continue his humanitarian efforts wherever they're needed. "Our guys really go deep," says Andrés. "We go deep to every corner...to make sure we reach those forgotten areas and forgotten people, and that's what makes it really different...We show up every day and we stay until the immediate emergency goes away."

JOSÉ ANDRÉS'
classic gazpacho

prep time: 15 minutes plus time for chilling • *makes:* 6 servings

Gazpacho is so versatile and amazing that it deserves to be a verb instead of a noun! Will it gazpacho? Whether fruit or vegetable, as long as it's juicy and soft, the answer is probably yes. At my home we gazpacho everything from asparagus to zucchini. Here's how you can do the same. *josé*

ingredients

base ingredients

2 cups crustless bread torn into 1-inch pieces
¼ cup sherry vinegar
¾ cup extra virgin olive oil
kosher salt and freshly ground black pepper

classic ingredients

1 cucumber, peeled, halved lengthwise,
 seeded and chopped
1 green bell pepper, cored, seeded, and diced
3 pounds plum tomatoes, quartered
2 garlic cloves
½ cup Oloroso sherry

topping suggestions

halved cherry tomatoes
croutons

Seasonal ingredients can be substituted for classic ingredients. Here is another variation, with seasonal melon added to base ingredients:

seasonal ingredients

Melon Gazpacho
1 tablespoon mint leaves
1 pound honeydew or other green melon,
 peeled and roughly chopped
1 green bell pepper, cored, seeded, and chopped
1 garlic clove
1½ cups water

topping suggestion

mint leaves

directions

1. Add classic ingredients to a blender, along with crustless bread and sherry vinegar. Blend until very smooth.
2. With blender running, slowly add olive oil, blending until emulsified. Pour mixture into a bowl and refrigerate until well chilled.

Recipe reprinted with permission from Vegetables Unleashed

slow cooker apple chickpea curry

pre-prep: If using dried chickpeas, soak 6 hours (or overnight) and cook for 1 hour (page 11)
prep time: 15 minutes • *cook time*: 4–5 hours • *serves*: 4–5

The combination of the deep, rich curry flavor combined with the sweet and tart green apples make this meal one of the tastiest curry dishes ever. The chickpeas not only add a nice texture, but also provide protein and fiber. The beauty of the slow cooker is that you can start it and forget about it until you are ready to eat!

ingredients

1½ cups coconut milk, canned

¼ cup coconut flakes, unsweetened

½ tablespoon olive oil

3 cloves garlic, minced

1 cup chickpeas, cooked and drained,
   or 1 can chickpeas, drained

1 onion, chopped

2 green apples, chopped

1 sweet potato, chopped into bite-size cubes

1 bell pepper, any color

1 teaspoon salt

1 tablespoon curry powder

1 teaspoon turmeric

½ teaspoon cumin

topping ideas

bunch of cilantro • lime slices

directions

1. Blend coconut milk with coconut flakes in a blender until smooth. Set aside.
2. In a small pan, heat olive oil on medium and sauté minced garlic for 3–4 minutes. Remove from heat and set aside.
3. Add coconut milk, garlic, and all other ingredients except cilantro and lime to slow cooker and cook for 4–5 hours on low.
4. When curry is done, serve over rice, squeeze some lime on top, and garnish with cilantro. Leftovers can be refrigerated.

dean
WILHELM

Dean Wilhelm is a Hawaiian kalo (taro) farmer who co-founded, with his wife Michele, Hoʻokuaʻāina, a nonprofit that teaches life strategies to at-risk youth through ancient Hawaiian traditions of kalo cultivation.

As a former Department of Education teacher at the Hawaiʻi Youth Correctional Facility and a Hawaiian musician, Dean realized that bringing Hawaiian food and culture into the curriculum was key to getting students more engaged. This, paired with Dean and Michele's desire to create a gathering place for people to connect with ʻāina (land) led them on their journey to find land in 2007 and officially create Hoʻokuaʻāina in 2011.

According to Dean, they never set out to become kalo (taro) farmers. "We're in the people business first and foremost," he says. "It's the growing of kalo that is our means to growing young people and community."

And grow a community they did. Dean worked as a teacher for the Hawai'i Youth Correctional Facility, but immediately noticed that he was having difficulty connecting with his students. Many came from challenging backgrounds, and it was a struggle to engage with, let alone teach them. After giving it some thought, he planted kalo right outside of his classroom as a way to connect with these young Hawaiians (75% of the student population) and create community.

Kalo is not only a highly nutritious root vegetable, but is also considered to be an ancestor of Hawaiian people. It is the oldest cultivated crop and arguably the foundation of Hawaiian society's history, as it enabled a culture isolated in the middle of the Pacific Ocean to thrive. In short, kalo is considered sacred.

"I believe in caring for a place... stewarding over it, and having the faith and confidence that it's going to care for me."

As these plants grew just outside of Dean's classroom, the students became intrigued.

Dean's class began to harvest the leaves of the kalo with their own hands and learned how to prepare laulau, a native Hawaiian dish wrapped in taro leaves. The majority of his students had never created this dish before and they took great pride in growing, preparing, and sharing this food. The real breakthrough happened when students started sharing their knowledge with other teachers and students.

At this point, Dean realized that this land and food had a healing power. The land became not only a place to gather in community, but also a place to work and create in community. The students learned much more than kalo cultivation, as this common ground allowed Dean to teach life skills to his students.

It turns out this was only the beginning. Dean and Michele realized that to make the most impact in these students' lives, they must do something more than just garden outside the classroom. They formulated a plan, sold their house, and in 2003, embarked on a journey to find land. In 2007, they found a beautiful parcel perfectly suited to create a gathering place and created Ho'okua'āina, which became an official nonprofit in 2011.

The name Ho'okua'āina was given to the organization by Kupuna (elder) Earl Kawa'a. It refers to their dedication to:

1. Keep the traditions of our kupuna alive by restoring 'āina back to abundance.

2. Build community through the sharing of traditions.

3. Pass on ancient knowledge to future generations.

Ho'okua'āina is located in Kapalai, in the ahupua'a of Kailua in Maunawili on the island Oahu. The journey to create the beautiful lo'i (water taro patch) that stands strong today began in 2002.

Hoʻokuaʻāina's mission is to empower youth to realize the meaning and purpose of their lives by helping them develop life strategies and skills through the cultivation of kalo and Hawaiian cultural values-based coaching. By growing kalo, this team empowers the next generation and builds stronger communities. "Not thinking in community is dated. It seems archaic," says Dean. He thinks the individualistic mindset has to go in order for everyone to thrive.

The Wilhelms and their team are always intentional about understanding why people come volunteer or visit the organization. They believe strongly that each person's life has meaning, value, and purpose, and the loʻi kalo (taro farm) at Hoʻokuaʻāina is a safe place to cultivate that. In Dean's own words, "You can't have well-being with an empty stomach ... You have to be at least satisfied, then the possibilities spring forth."

The reliance on traditional Hawaiian agriculture methods at Hoʻokuaʻāina honors the land and also inspires others. Now thousands of people from all facets of the community come to the farm annually to participate in the programs and help to cultivate kalo. They are the backbone of the land, and the land is their backbone.

"I always pondered 'how did people live ... and survive [back then]?' Kuaʻaina literally means back land, the backbone of the land. It refers to the people who are the backbone of the land and actively live the Hawaiian culture to keep the spirit of the land alive."

Dear Dad and Grandpa,

I never had the chance to thank you both for sharing the gift of cooking with me. I was destined to be a "foodie," having a dad who was a professionally trained chef of the highest caliber from Switzerland and a Hawaiian/Chinese grandpa who owned a restaurant and often catered for up to a thousand people Hawaiian style. Looking back, my life experience of eating and partaking in the cooking of such an amazingly broad spectrum of food that I took as normal was incredible and truly unique. Thank you both for all the meals you cooked for me and for taking the time to share with me the basics of cooking that I use to this day. What I would give to watch both of you cook again! Even more so, thank you for inspiring me so that cooking is my most creative outlet and passion. Writing this letter has really made me miss you both. A hui hou aku no i kuke pū kākou i ka lani! Me ke aloha pau'ole.

dean

DEAN WILHELM'S
'āina curry

prep time: 1½ hours (including the steaming of kalo and ulu)
cook time: 20 minutes • **makes:** 8 healthy servings

ingredients

2 cups kalo (*taro*), cubed

2 cups ulu (*breadfruit*), cubed

1 medium-size onion, chopped

3 tablespoons avocado oil

2 tablespoons Thai curry paste

4 tablespoons fresh grated 'ōlena (*turmeric*)
(*If dry then 1 tablespoon*)

5 cloves garlic, chopped

32 ounces coconut milk, canned

3 average-sized Japanese eggplant cut into 1" cubes

1 bunch kale, chopped

1 lime

Hawaiian salt to taste

directions

1. Steam the kalo and ulu for 45 minutes in a pressure cooker or a little longer in a steamer. Cook until fork tender.

2. Sauté chopped onion with avocado oil, then add curry paste, grated 'ōlena, and chopped garlic. Continue to sauté for 5 minutes.

3. Add coconut milk to sautéed spices and onion. Let simmer for a few minutes, then add eggplant. Cover and simmer for 2 minutes.

4. Add in steamed kalo and ulu. Cover and simmer for another 5 minutes.

5. Turn off heat. Stir in chopped kale and the juice of one lime. Salt to taste. Let sit for 5 minutes. The kalo and ulu will soak up a lot of the liquid. Add a little water or chicken broth if too thick.

6. E 'ai! (then eat!) I mā'ona ka 'ōpū (until your stomach is content).

sweet potato gnocchi

prep time: 40 minutes • **bake time:** 1 hour plus 10 minutes of letting the dough sit • **serves:** 4–5

At first, we were intimidated by the thought of making gnocchi. Whether it was the fancy name or the rich history of this Italian staple, we had the impression that it wasn't something we could make at home. Turns out, it is super simple to make and uses very few ingredients. While a typical gnocchi recipe would involve a plain potato and wheat flour, this recipe calls for sweet potatoes and gluten-free flour, making it higher in fiber and vitamin A, as well as accessible for those who are gluten-free! Try using our Marinara Sauce (page 178) or Alfredo Sauce (page 180) with this pasta.

ingredients

2 large sweet potatoes
2 cups almond flour
1¼ cup arrowroot powder + extra for rolling

1 teaspoon salt
1 teaspoon garlic powder
1 flax egg (*page 10*)

directions

1. Preheat oven to 350°F.
2. Pierce potatoes with a fork and bake on a baking sheet until tender, about 1 hour.
3. In a large bowl, combine almond flour, arrowroot powder, salt, and garlic powder. Set aside.
4. Once potatoes are done (no need to remove the skin), mash with a fork until smooth. Mix in the large bowl with other dry ingredients. Mix in egg. Using your hands, knead the mixture into dough. If dough is sticky, dust hands with almond flour or arrowroot powder.
5. Form the dough into a ball and cut into 8 even-sized portions. Let dough sit for 10 minutes.
6. Lightly dust a flat working area with arrowroot powder. Roll each portion into a rope that is ½-inch wide. Using a knife, cut each rope into 1" pieces, then form into balls and lightly dust each one with the powder. If the balls are dry, use your hands to lightly coat them with a small amount of coconut oil.
7. Using a fork turned upside down, roll each gnocchi across the lines of the fork and press down gently. The indentations help the gnocchi hold the sauce better and allow them to cook faster.
8. In a medium pot, bring 3 quarts of lightly salted water to a boil. Drop the gnocchi balls into the pot, about 15–20 at a time. The gnocchi are done when they float to the surface and remain there for 15 seconds. Using a slotted spoon or skimmer, remove the gnocchi from the water and shake off excess water.
9. Place the cooked gnocchi on a lightly greased baking sheet (coconut or almond oil is fine), which will help prevent them from sticking together.
10. Toss the gnocchi with Alfredo Sauce (page 180) or Marinara Sauce (page 178).

sweet potato enchiladas

pre-prep: If using dried beans, soak 6 hours or overnight and cook for 1 hour (page 11),
If making Cheesy Cheddar Sauce, 10 minutes (page 181),
If making Cassava Flour Tortillas, 20 minutes (page 149)
prep time: 55 minutes • cook time: 25–30 minutes • serves: 5–6

If you're looking for a meal to wow your meat-eating friends (and your vegan friends as well), this recipe is for you. One of our most colorful dishes, these enchiladas' sweet potatoes and black beans make them incredibly filling. Yet, unlike classic enchiladas that are loaded with cheese and leave you feeling stuffed, this dish makes you feel nourished. Enjoy it with our Cauli/Cashew Sour Cream (page 186).

ingredients

homemade enchilada sauce

1 tablespoon olive oil

1 tablespoon gluten-free flour

2 tablespoons chili powder

½ teaspoon garlic powder

¼ teaspoon cumin powder

¼ teaspoon oregano

1½ cups vegetable stock

1 tablespoon tomato paste

salt to taste

filling

1 tablespoon olive oil

1 onion, diced

¼ teaspoon salt

1 red bell pepper, diced

1 large sweet potato, cut into small cubes

3 garlic cloves, minced

½ cup crushed tomatoes

1 tablespoon chipotle paste (optional) or
    1 teaspoon chipotle seasoning

1½ cups beans, black or navy, cooked and drained

6 or 7 Cassava Flour Tortillas (page 149)
    If you don't want to make the tortillas yourself, try any type of tortilla from the Siete Foods brand!

1 cup Cheesy Cheddar Sauce (page 181)

Recipe continued on page 114

topping ideas

avocado • Cauli/Cashew Sour Cream *(page 186)* • cilantro or coriander

directions

enchilada sauce

1. In a small pot, warm olive oil. Add flour and dry seasonings and stir together. Add vegetable stock and tomato paste, stir until smooth. Let sauce simmer for 7–10 minutes. Remove from heat and set aside.

enchiladas

1. Preheat oven to 400°F. Lightly grease 9x13" casserole dish.
2. Add oil to a large skillet and warm to medium heat. Add onions and salt and sauté for 4–5 minutes. Add red bell peppers, sweet potato, and garlic. Stir for a few minutes, until potatoes start to soften.
3. Add tomatoes and chipotle paste or seasoning to skillet and stir to coat the veggies. Cook for a few minutes. Add cooked beans and mix. Remove skillet from heat.
4. Pour ½ cup of the enchilada sauce into a casserole dish, and spread evenly to coat bottom.
5. Assemble enchiladas by laying 1 tortilla down on a plate and spooning 1/3 cup of the sweet potato mixture in the middle of the tortilla. Wrap sides of the tortilla over the filling. Continue to do this for the next 4–5 tortillas. Place stuffed tortillas folded-side down in casserole dish.
6. Pour remaining enchilada sauce over top of tortillas. Sprinkle or drizzle the cheese over the top. Bake for 25 minutes. For extra-golden-brown tortillas, cook an extra 4–5 minutes.
7. Remove from oven and let cool. Add suggested toppings.

sloppy joes

pre-prep: 30 minutes to make Lentil Walnut Meatless Meat (page 170), 20 minutes to make Mozzarella Cheese (page 184), 30 minutes to make Marinara Sauce (page 178), 1 hour and 45 minutes to make Gluten-Free Burger Buns (page 140)

prep time: 10 minutes • *makes*: 4 servings

This recipe will surely send you back to your school cafeteria or sleepaway camp days. It not only tastes incredible, but also encourages people to be as messy as they want. We also love the fact that you will be able to put four of our recipes together to create this. If you plan ahead, you can refrigerate those four dishes, then combine them all for serving. Don't forget the napkins!

ingredients

2 cups Lentil Walnut Meatless Meat (*page 170*)
1 cup Mozzarella Cheese (*page 184*)
2 cups Marinara Sauce (*page 178*)
4 Gluten-Free Burger Buns (*page 140*)

directions

1. Toast hamburger buns (there should be 8 halves).
2. Add ¼ cup Lentil Walnut Meatless Meat on each bottom half.
3. Add ¼ cup of Marinara Sauce on top of each half.
4. Top with 1–2 tablespoons Mozzarella Cheese. Enjoy!

chickpea "tuna" salad sandwich

pre-prep: *If using dried chickpeas, soak for 6 hours or overnight and cook for 1 hour (page 11)*

prep time: *15 minutes* • *makes*: *4 sandwiches*

My love for tuna sandwiches definitely came from my dad. A simple tuna sandwich is his favorite meal in the world. Every Saturday he would pack tuna sandwiches and we would head to the beach for my weekend beach volleyball tournaments. Then, while I competed in volleyball, my dad would have his own competition: comparing his tuna sandwich creation with those made by the other dads. They would try to outcompete each other with superior tuna salad, bread, pickles, and tomatoes. I think they might have been more interested in the sandwich competition than the volleyball games! When writing this book, it was important for me to create a sandwich that reminded me of these times, but also to use ingredients that are better for our bodies and for the planet.

And now for the not-so-fun facts: As tuna are among the world's most popular fish, overfishing is a huge issue. Pacific bluefin tuna have been overfished to near extinction, and in recent years, the Atlantic bigeye and the Indian Ocean yellowfin tuna have come to suffer from overfishing, as well. Many marine populations depend on tuna as their primary source of protein. If the the tuna fisheries are not regulated correctly, this will become an ecological disaster for our oceans. Futhermore, although tuna is a healthy food alone, they contain high levels of mercury due to water contamination (likely from coal-burning power plants). High exposure to mercury can disrupt brain function and harm the nervous system. All that to say, we're glad you're here to try this recipe! *mackenzie*

ingredients

1 cup chickpeas, cooked or canned, drained

3 tablespoons tahini

1 tablespoon vegan mayonnaise

1 teaspoon Dijon mustard

1 tablespoon maple syrup

¼ cup onion, diced

¼ cup celery, diced

¼ cup pickle, diced

pinch of salt and pepper

directions

1. Place chickpeas in a food processor and pulse about 8 times until chickpeas have a tuna-like texture.

2. Add the rest of the ingredients to the bowl, and mix.

3. Enjoy as a sandwich or on a bed of fresh garden greens.

maricela VEGA

With over a decade of professional cooking experience and a background in agroecology, Maricela Vega is a food justice advocate striving to get to know local farmers and to tell their stories through her food.

Vega received a 2018 Eater Award and a 2020 James Beard nomination for her work as a chef. Most recently, she was the executive chef at 8ARM, now a Japanese-forward restaurant in Atlanta, Georgia. She is currently focusing on launching her business, CHICO, a "masa goods y mas" ["masa" is dough made from corn flour; "y mas" means "and more"] social enterprise based in Atlanta.

If you have the good fortune to meet Maricela Vega (known as Mari), you will learn right away that she is a very curious person. The first time we met her, we were on a bus in Cuba. We were all waiting for one expected passenger, Vega, who had stopped to quickly climb a tree and forage some fruits she saw. It is no wonder that she is so knowledgeable about the world around her.

Vega's upbringing had a profound impact on her cooking. Her parents always maintained a garden with various herbs and vegetables. "They always had a little something growing everywhere they lived … agave, mint, some sort of little plant. They were always saving tomato or pepper seeds." She began to connect the dots between fresh food and the ability to produce incredibly tasty meals. She appreciated the story and origin of a dish's ingredients. Taste and story, to Vega, are of equal importance.

Raised in the South (North Georgia), she grew up eating traditional Mexican home cooking made daily by her mother. Vega is a second generation Mexican American whose ancestors were maize farmers. Her grandparents had a farm in Guanajuato, Mexico, and her family would visit as often as possible. At the farm, she came to appreciate her grandmother's cooking and spent time watching her grandfather and brothers work on the land. Although she was too young to harvest, she dreamed of someday being able to grow that delicious food. Visiting the farm made Vega understand the hard work that goes into producing good food and allowed her to taste the beautiful flavors of Mexico from a young age.

On one of her visits to Guanajuato, Vega stood on the mesa looking out at the farm. She imagined her whole family up there and decided that she needed to learn more about food and farming.

During a period of education and a criminal justice internship in Atlanta, Vega took several breaks from the legal world. She started cooking, and eventually joined a veganic (vegan + organic) farm, Grow Where You Are. This move fueled her desire to impact Southern food culture. After a two-and-a-half year agroecology apprenticeship, she became a chef in order to teach Atlanta consumers about authentic Mexican cuisine and culture.

During a decade of food-related work, Vega has conducted research about maize and nixtamalization, the process for preparing maize, in which the corn is soaked and cooked in an alkaline solution (usually limewater), washed, and then hulled. She also worked as an executive chef, developed recipes, and was a prop maker and food styling specialist. Being involved in the restaurant scene made her question where the food she was prepping came from.

Visiting the farm made Vega understand the hard work that goes into producing good food and allowed her to taste the beautiful flavors of Mexico from a young age.

Dear Ma Diega and to all the grandmothers,

Hola Ma Diega... the last time I saw you it was for a week in the spring of 2017, just as the earth was beginning to awaken and there were so many splashes of flowers lighting up the landscape. Despite the short visit, it felt like a very long week and somehow you managed to gift me the process of making candied calabaza (squash).

A pile of hillside-foraged squashes lay in your patio; curious, I asked, "What to do with all of these, Ma!?" Your bright smile and eyes lit up the room as you confidently answered, "dulces (candy)."

I wheeled you out into the patio and asked for your guidance as I peeled away the skins and methodologically poked holes throughout the squashes, preparing them for a 24-hour nixtamal solution soak followed by cooking over an open 12-hour firewood flame. You were always such a talented cook, flavors so delicate, techniques so carefully perfected with time.

Last summer, in 2021, I returned to your home, rather our home, for the first time since you left this earth. The fields were luscious and the rains were about to begin. As I pulled in, it felt so physically empty, but yet you were felt. In fact, you are always felt. That week we found you in every dish my tias (aunts) and I made/ate. Your presence was one with our assembly and procedure. I carry you and our lineage daily through the foods I prepare. You, like many women of the kitchen or households before me, inspire me daily. It is through your efforts of preservation that I am able to continue this practice.

Thank you for such invaluable knowledge, a true gift.

maricela

For example, she observed and tasted a huge difference between the peppers that arrived from a truck versus the pasillas and chiles negros peppers grown in San Jose de las Pilas, the town in Guanajuato where her grandparents' farm is located. She understood that when food travels many miles, this not only raises ethical and environmental issues, but also compromises taste. Her curiosity led to further questions: "What is your soil nutrition, what are your nitrogen fixers, why does this taste so good?" Her focus is on the growers, and she uses her experience to deconstruct, then try to improve the way food is grown and transported in the South.

Today, Vega is a well-known Atlanta-based chef redefining Mexican-American food in the South. She combines the recipes of her Mexican ancestors with local Southern produce. She believes it is her responsibility to cultivate relationships with both farmers and distributors, especially Indigenous ones. Using local seasonal ingredients also sparks Vega's creativity and innovation.

Social justice is always at the heart of Vega's work. During the early stages of the pandemic, when she was the executive chef at 8ARM, she and her team transformed the restaurant into a CSA and teamed up with the Living Walls nonprofit to ensure that for every CSA box purchased, another would go to an undocumented family through Freedom University.

The undocumented students reported back how grateful they were to have fresh local produce, and how different it tasted. Vega says in an interview with Zagat, "I'd estimate that roughly 80% of the entire food sector in the U.S., from seed, harvesting, selling, packaging, processing, and preparing ... is thanks to the work of undocumented immigrants ... I've heard of people who only make $15

a day. [Factoring] in how much these folks have been carrying on their backs—the weight, literally and figuratively, ... shows that we need to change the inequality in all of this regardless of their status."

When asked about her mission, Vega says she's trying to, "preserve and create connections, and all of that with intentionality. If anything, I know we can produce ... food that is delicious." Her goal is to produce food in the way her ancestors did. Long term, she'd like to return to her family's land in Mexico and revitalize that land to grow at least half of the produce for her business, CHICO. There's nothing like the taste of local peppers!

Vega adds that, "CHICO is focused on my ancestral roots ... creating Mexican provisions ... It can be looked at as honoring the contributions that MesoAmericans have made through the nixtamal and agriculture. How amazing it is to have a 10,000-year-old process that people still use." Vega says she wants to center more on that narrative and show people the different versions of what nixtamal can look like, through the tamales, tostadas, sopes, and tlacoyos. She is set on decolonizing Mexican food, providing access to it, and embracing vegetables in her cooking.

MARICELA VEGA'S
carolina gold rice with salsa macha and cilantro

prep time: 25 minutes • **cook time:** 20 minutes
makes: Approximately 3 cups, 1 tablespoon per serving / 16 servings per cup

This recipe is simple to make, but it's best to prepare the salsa macha in advance. The salsa is oil based, so it maintains a great shelf life. A splash of very good soy sauce and citrus will tie it all together.

ingredients

1 cup rice
2 cups water

salsa macha

2½ cups grapeseed oil
¼ cup minced garlic
2 ounces morita pepper
2 ounces pasilla pepper
2 ounces guajillo pepper
¼ cup sesame or benne seeds
¼ cup pumpkin seeds
1 teaspoon black pepper, ground
1 teaspoon cumin, ground
1 teaspoon coriander, ground
1 teaspoon good salt
1 teaspoon MSG *(optional)*
1–2 teaspoons soy sauce
1–2 teaspoons citrus juice *(lemon or lime)*
cilantro, dill, chives, or basil *(optional)*

directions

1. In a pot, add rice to water and cook for 15 minutes or so on low heat.

2. Set a sauce pot over medium heat and add the oil. Once the oil is hot, but not smoking, add the garlic cloves. Stir and fry for about 2 minutes or until they are golden, then remove garlic from the oil and place it in a large sauce vessel.

3. Repeat this process in batches in the following order: Add the chiles, then remove them into the vessel; then move onto the seeds (each time frying the components for about 2 minutes or until the bubbles subdue, indicating that moisture has been mostly removed, which allows for the crispy texture).

4. Finally, add the spices and shut off the heat. Allow them to cook in the residual heat, stir, and fry for about two minutes.

5. Carefully transfer all the contents into the jar of a blender and blend into a coarse grind, or blend longer for a finer texture. Let cool for about 10 minutes. If you prefer, you may leave the seeds whole for more texture.

6. To assemble: Scoop a cup of rice into a bowl, layer in a good spoonful of salsa macha, a few teaspoons of soy sauce, and citrus juice such as Meyer lemons or limes. Then finish with cilantro or other herbs such as dill, chives, or basil. If you have a tomato, feel free to add it in.

Vega combines the recipes of her Mexican ancestors with local Southern produce.

Factory Farms

Imagine living your entire life in a cage the length and width of your body—a cage so small that you're unable to turn around. This is the harsh reality for 97 out of 100 pigs raised in the U.S., where factory farms or CAFOs (Concentrated Animal Feeding Operations) have become the norm. In these indoor "farms," large numbers of livestock are raised in deplorable conditions, for the sole intention of maximizing production at minimal cost. Industrial farming has become so commonplace that 90% of all global meat—as well as 99% of meat in the U.S.—comes from these facilities.

Like pigs, most chickens also spend their short lives under appalling conditions. The nonprofit Mercy for Animals reports that in the U.S. alone, 99.9% of the nearly 9 billion chickens killed annually for their meat. These buildings are packed with birds, making the ground invisible. In a natural setting, chickens live for five to eight years. In CAFOs, they get 47 days in a dusty building, unable to spread their wings.

Yet, many consumers don't know that their meat and dairy come from these massive industrial operations. Packaging for milk and other dairy products features images of cows roaming freely in country fields. Childhood depictions of animal farms show chickens and cows on bright green grass and pigs rolling playfully in the mud. Tragically, factory farms are far from this reality—many animals on these farms never see sunlight or take a breath of fresh air.

How is this reality hidden from so many of us? Why do we still think of farming as what we see on the packages? The lack of information is, in part, due to what's referred to as "ag-gag laws," which prohibit audio or visual recording at industrialized farming operations. Twenty-five states have attempted to pass modern-day ag-gag laws, and six

of those states (Iowa, Utah, Missouri, Idaho, Wyoming, and North Carolina) have succeeded. The first five impose criminal penalties, while North Carolina's is the first in the nation to impose a civil sanction. I (Mackenzie) spoke with *National Geographic* photographer George Steinmetz, who described being arrested after taking photos over a feedlot, even though there were no warning signs.

Even in places other than the states with ag-gag laws, the public is largely uninformed about the conditions animals endure. For example, a UK-based animal rights group conducted a study and found the following:

- 33% of adults didn't know dairy cows were slaughtered when their milk yield dropped. Instead, many thought cows retired to animal sanctuaries.

- Similarly, 88% didn't know that most pigs were killed at just six months of age, even though their natural life span is 15 years.

- Two-thirds of those in the study were unaware that killing all male chicks on egg farms was standard practice. About 69% stated the practice should be illegal.

- After learning about factory farming, nearly half said they would consider cutting back on animal products. Some people (16%) even said they would give up meat and dairy entirely.

In addition to the harm they inflict on animals, factory farms are especially toxic to those who live near these facilities. The waste is often stored in massive open-air cesspools, and workers are allowed to spray this waste on surrounding lands that are usually near people's homes. The toxins and harmful gasses released cause respiratory infections, antibiotic-resistant superbugs, and other illnesses. These operations are generally located in low-income areas, where people often can't afford to move and are forced to suffer the negative health impacts.

These facilities also produce greenhouse gasses such as methane, which is 25 times more harmful than CO_2, that circulate in the atmosphere, contribute to climate change, and pollute the environment. Animal agriculture is the second-largest contributor to human-made greenhouse gas emissions after fossil fuels and is a leading cause of deforestation, water and air pollution, and biodiversity loss. According to the group Food and Water Watch, pollution and waste from industrial agriculture threaten over 60,000 acres of lakes and ponds and over 13,000 miles of rivers and streams.

How is this reality hidden from so many of us?

Whether you're a vegan, a meat eater, or anything in between, the conditions of factory farms are indisputable and distressing. According to a nationwide survey commissioned by the ASPCA in August 2020, "The vast majority (89%) of Americans are concerned about industrial animal agriculture, citing animal welfare, worker safety, or public health risks as a concern." Yet, in our current food system, 10 billion land-based animals are raised for food each year in the U.S., and the overwhelming majority of them live in conditions that do not align with these humane customer values. How can we create change? In addition to modifying our own diets, we can educate ourselves about the conditions on factory farms and then support and vote for the implementation of welfare regulations against these facilities. Groups like Food and Water Watch provide action steps you can take to make a difference.

See page 230 for a list of resources for how you can get involved and help!

Did you know that it takes 660 gallons of water to produce *one hamburger?* If you're reading this, you are at least somewhat committed to trying an alternative. So if you're looking for a hearty, delicious veggie burger to fill you up, we got you. We couldn't pick just one, so here are *two awesome burger varieties* to choose from: Sweet Potato Burger and Black Bean Beet Burger. Serve the burger with our toasted Gluten-Free Burger Buns (page 140).

black bean beet burger

pre-prep: If using dried black beans, soak for 6 hours or overnight and cook for 1 hour (page 11). Green lentils: cook for 45 minutes (page 11) • **prep time**: 30 minutes • **cook time**: 20 minutes • **makes**: 6 patties

This delicious burger is packed with protein (shout-out to beans, lentils, pumpkin seeds, and quinoa flour) and has incredible flavor thanks to the combination of spices. The mushrooms add an excellent meaty texture, making this burger our favorite meat substitute.

ingredients

⅔ cup pumpkin seeds

½ cup oats

1 cup quinoa flour

1 medium beet

1 tablespoon olive oil

1 cup mushrooms, sliced

½ cup yellow onion, diced

4 cloves garlic, minced

½ teaspoon apple cider vinegar

½ teaspoon cayenne pepper

½ teaspoon smoked paprika

½ teaspoon ancho chile powder

½ teaspoon curry powder

½ teaspoon garlic powder

½ teaspoon mustard powder

½ teaspoon ground white pepper

½ teaspoon black pepper

½ teaspoon salt

1 cup black beans, cooked or canned, drained

(from either method, set aside ¼ liquid from beans)

1 cup green lentils, cooked

2 tablespoons oil for frying (olive, avocado, or coconut)

topping ideas

Cheesy Cheddar Sauce (page 181) • Thousand Island Dressing (page 189)

tomato • onion • lettuce • avocado

directions

1. Add pumpkin seeds and oats to blender and blend into a flour. Add mixture to a large bowl along with quinoa flour. Mix well and set aside.

2. In a small pot, steam beet with steamer basket for 10 minutes, until beet is tender. Remove from heat and let cool. Peel skin off and slice.

Recipe continued on page 128

3. Heat olive oil in large skillet over medium-high heat. Add beet, mushrooms, onions, garlic, and apple cider vinegar. Combine spices in a small dish and add to vegetables. Sauté for 10–15 minutes.

4. Add cooked veggies to a food processor, along with beans, liquid from beans, and lentils. Pulse about 10 times, until everything is combined but still has some texture.

5. Add veggie mixture to the large bowl with flour mixture and mix well.

6. Form patties using ⅓ cup mixture.

7. Add frying oil to pan. Heat pan to medium heat and fry 3 patties at a time for 3–5 minutes on each side until lightly browned.

8. Serve the patties on toasted Gluten-Free Burger Buns (page 140) with toppings of your choice.

sweet potato burgers

pre-prep: 1 hour (to bake sweet potatoes), 20 minutes (to cook quinoa and lentils)

prep time: 15 minutes • **bake time**: 20 minutes • **Frying time**: 10 minutes • **makes**: 9–10 patties

*T*hese burgers are the perfect blend of sweet and savory, and sweet potatoes are a good source of fiber and potassium. To top it off, the combination of sweet potatoes with the quinoa, lentils, tahini, and nut butter is not only nutritious but creates an excellent binder that holds the burger together.

ingredients

2 large sweet potatoes

1 cup quinoa, cooked

1 cup lentils, cooked (*any kind*)

1 cup oat flour

2 tablespoons nut butter (*almond, peanut, or cashew*)

2 tablespoons tahini

2 tablespoons maple syrup

2 tablespoons nutritional yeast

1 teaspoon garlic powder

1 teaspoon salt

¼ teaspoon pepper

3 tablespoons coconut oil for frying

topping ideas

Cheesy Cheddar Sauce (*page 181*) • Thousand Island Dressing (*page 189*)

tomato • onion • lettuce • avocado

directions

1. Preheat the oven to 375°F.
2. Poke a few holes in the sweet potatoes using a fork.
3. Bake sweet potatoes for 1 hour. After potatoes are done, let the oven remain on.
4. Meanwhile, add 1 cup of quinoa to 2 cups water in a medium saucepan. Bring to a boil, and simmer for 15 minutes.
5. Next, in another saucepan, add 1 cup of lentils to 2 cups water. Bring to a boil and simmer for 20 minutes.
6. When sweet potatoes have cooled down, cut them into 4 pieces each and mash with a fork. Leave skin on for extra nutrients.
7. In a large bowl, add mashed sweet potato plus all other ingredients. Use a fork to combine and mash together.
8. After mixture is blended, form 10 patties.
9. Line baking pan with parchment paper. Place patties on a baking pan and bake for 20 minutes.
10. After patties are done, use a medium/large frying pan with 1 tablespoon coconut oil to lightly brown 3 patties at a time. Cook on each side for 3–4 minutes.
11. Serve the sweet potato patties on toasted Gluten-Free Burger Buns (page 140) with toppings of your choice.

sides

cornbread

prep time: *15 minutes* • **cook time:** *30 minutes* • **makes:** *12 pieces*

Cornbread is the perfect addition to Dad's Hearty Chili (page 68) or Sweet Potato Harissa Soup (page 162), or delicious on its own with some farm-fresh jam. Most cornbread recipes go heavy on the butter, milk, eggs, and granulated sugar. With this recipe, we have found a way to make an awesome tasting cornbread without the animal products.

ingredients

1 cup gluten-free flour mix

1 cup cornmeal

1 tablespoon baking powder

¼ cup coconut sugar

1 teaspoon salt

1 cup almond milk

¼ cup coconut oil, melted

3 flax eggs *(page 10)*

¼ cup honey

directions

1. Preheat oven to 400°F. Line the inside of an 8x8" pan with parchment paper or coat pan with melted coconut oil or coconut spray.

2. In a large bowl, mix the flour, cornmeal, baking powder, coconut sugar, and salt. Mix together and set aside.

3. In a medium bowl, combine almond milk, melted coconut oil, flax eggs, and honey. Mix well.

4. Pour wet ingredients into large flour bowl and mix with a hand mixer for 1 minute.

5. Pour mixture into prepared pan, bake for 25–30 minutes. Let cool before slicing.

DR. *gail* MYERS

Dr. Gail Myers is a cultural anthropologist and the founder of Farms to Grow, a nonprofit dedicated to working with Black farmers and other underserved sustainable farmers around the country.

Farms to Grow initiated the creation of Oakland's Freedom Farmers' Market with the mission of bringing traditional legacy foods from Black and other sustainable farmers into Oakland. Myers has conducted field research, lectured, written about, and filmed stories of African American farmers, sharecroppers, and gardeners for over two decades and is considered an expert in African American farming knowledge, traditions, and adaptations. Her upcoming multimedia documentary, *Rhythms of the Land*, which is an ode to Black farmers in the United States from the enslavement period to the present, is currently in post-production.

The Power of Women: "Most of the world's farmers are women. And they are losing land. The role of women is so important. Women save the seeds, women have the nurturing. The earth is a womb, I believe. We are natural stewards, huggers of this earth. We are protectors of it. And so if women can't farm and aren't on the land, we are missing a whole other level of nurturing and our survival strategies to be on this planet."

Dr. Gail Myers' love for nature dates back to her childhood days in Florida, where she lived near the ocean and connected with the Earth by swimming every chance she had. She came from a community that had a collective-vision perspective, which helped to shape her life's work. From a young age, she was taught that, "if you were mentored within your community, and you were given the opportunity to go out and get extra, it was expected that you would give part of what you earned to a communal fund because it was the community that got you there." Myers has dedicated herself to educating communities about the rich historical traditions of Black farmers. She also takes part in a wide spectrum of grassroots organizing and coalition building through her work with Farms to Grow.

To gather material for her film, *Rhythms of the Land*, Myers drove 10,000 miles in four weeks in the summer of 2012. She visited 10 southern states (South Carolina, North Carolina, Georgia, Tennessee, Texas, Alabama, Mississippi, Arkansas, Louisiana, and Florida), interviewing over 30 African American sharecroppers, tenant farmers, and third-to-fifth-generation farmers.

The film spotlights an array of vegetable farmers, rice growers, hog ranchers, dairy ranchers, barefoot farmers, sharecroppers, basketweavers, shrimp farmers, and gardeners, each sharing their history. The interviews represent generations of cultural traditions, family farming, and a philosophy that honors sustainability and community.

"I am out here with these farmers, walking the land and listening to their revelations," Myers says. One of these revelations comes out in a story. "I was walking up a patch of fence with a farmer and he says, 'You know, Gail, I think the days are getting shorter. Me and my grandfather used to walk this land, and me and my father used to walk this land. I remember how long it would take, and now we walk this same stretch of fence and the sun sets sooner by the time we get to it.'"

Chef Wanda Blake cooked a meal for a scene in Rhythms of the Land.

Myers observes that, "these are people that watch every nuance of their world. They know a place-based way of reviving the ecological environment like no one else. Those relationships extend for hundreds of years. So, what I began to understand when I worked with these farmers … is that it was important that I was having this relationship not only with these folks, and these farmers and this food, but also with this land. I finally understood that this work was about freeing the land, and it was so crucial that those narratives were there to be part of a repertoire, a repository of community knowledge."

In one interview after the next, Black farmers and ranchers demonstrate their ongoing perseverance and commitment to the land despite predatory lending practices and policies levied against them.

"African American farmers have been ignored and denied any real place in the agricultural history and literature of the United States," Myers states on her website. "Although [they] have been virtually invisible from the stories of American agriculture, our African American farmers hold vast agricultural knowledge, some of which we now know as sustainable agriculture. This documentary film will reveal that knowledge and the experiences of African American farmers yesterday and today."

When asked about the overall mission of her work, Myers states, "I think it's just very simple. Leave this place better off than we find it and make it possible for life to continue in a way that flourishes so that we can see seven generations, ten generations ahead of us. Follow your heart, follow your passion."

To Mr. William Chambers and all farmers before him,

To those farmers who labored the land bound by chains of enslavement. To those who farmed beyond the whip because that is what they knew. To those who worked and eventually owned their land to keep our dreams alive. You all fought the good fight. You held this precious land in your hearts either by force or by hope, and you worked it with your hands. I see you, all of you. In spite of the tragedies of loss, our communities grew and generations were raised up on the food you all planted with hope. This letter is to say thank you for dreaming our dreams! We are here still farming the land, still working the dreams, and finding you all there with us too. Thank you. Thank you. I love you all!

gail

FROM DR. GAIL MYERS:

Ine's cornbread

prep time: 15 minutes • *cook time:* 45 minutes • *makes:* 15 pieces

Dear Mom, Growing up I remember you in the kitchen. You were always insisting that I watch you and learn how to cook. But I was always trying to escape the kitchen and your cooking lessons. It was just too hot in the kitchen for me. But I remember staying close to you on holidays and lending a hand for chopping vegetables and mixing the cakes and pies. However, there was one recipe I did pay attention to only because I loved it so much … cornbread. I watched and listened patiently as you taught me how to prepare cornbread from scratch. I thank you for those rare cooking lessons, Momma. You make the best cornbread from scratch. Thank you also for imparting to me the importance of education and learning. And thank you for teaching the joy of cooking from scratch.

ingredients

2 cups organic flour

2 cups organic cornmeal

2 teaspoons baking powder

½ teaspoon baking soda

¾ cup organic sugar

¼ cup olive oil

1½ cups milk

2 eggs

1 tablespoon butter or oil to grease cast-iron skillet

directions

1. Preheat oven to 350°F. Grease one 12-inch cast-iron skillet with butter or oil.
2. Sift the flour and cornmeal in a large mixing bowl. Combine all dry ingredients.
3. Add wet ingredients and stir mixture until combined and without lumps.
   Pour mixture into cast-iron skillet or two small baking pans.
   Bake for 30–35 minutes or until firm in the middle.

To make it plant-based:

• plant-based milk can be substituted for milk

• flax eggs can be substituted for eggs (page 10)

• vegan butter can be subtituted for butter

einkorn bread

prep time: 30 minutes • **rise time**: 1 hour, plus an additional 30 minutes after rise • **bake time**: 40 minutes

Because the two of us are both sensitive to modern-day wheat, we were so excited when we came across einkorn flour. Einkorn is an ancient wheat, and the only wheat in existence that has never been hybridized. Einkorn contains more protein than other grains, and because of its composition and DNA structure, can be easier to digest than modern wheat. We still suggest using careful judgment if you are sensitive to gluten; however, there are many success stories of people who suffer from gluten sensitivity but are able to tolerate einkorn flour. Plus, you'll love the bread's toasty, sweet, and nutty flavor. We suggest pairing this bread with Sweet Potato Harissa Soup (page 162), or enjoying it on its own with jam!

ingredients

4 cups einkorn all-purpose flour

1 cup water, warmed to 110–115°F

2¼ teaspoons active dry yeast

¾ cup water

¼ cup + 1 teaspoon honey

2 tablespoons olive oil

1 teaspoon salt

directions

1. Add flour to a large mixing bowl. Grease bread loaf pan with oil of choice. Set both aside.
2. Combine 1 cup water and 1 teaspoon honey in a pot and warm to 110–115°F. Add yeast and lightly mix. Let sit for 8–10 minutes until yeast proofs (foamy bubbles appear).
3. While waiting for yeast mixture, warm ¾ cup water, ¼ cup honey, olive oil, and salt to 110°F. When yeast mixture is ready, add both yeast and olive oil mixtures to the large mixing bowl with flour. Mix lightly until all ingredients are combined.
4. Cover bowl with a towel and set to rise in a warm area.
5. After 1 hour, coat your hands with a bit of extra flour and form dough into a loaf. The dough will be sticky, so add a little flour to be able to handle it. Do not knead or overwork the dough.
6. Place dough in greased loaf pan, cover with a towel, and let dough rest for 30 minutes. The dough will again rise. Preheat oven to 375°F.
7. After 30 minutes, remove towel and place bread in the center. Bake for 35–40 minutes, until a golden crust forms on top.
8. When bread is done, turn oven off and open door. Leave bread in oven for 5–10 minutes to prevent bread from sinking.
9. Allow bread to cool before slicing.

gluten-free burger buns

prep time: 20 minutes • rise time: 1 hour • bake time: 25 minutes • makes: 8 buns

We've got two awesome burger recipes, and now it's time to focus on the bun that holds it all together. Gone are the days when you swapped your fluffy, comforting bun for a lettuce wrap in an attempt to feel less bloated after eating a burger. The combination of alternative flours not only produces the perfect bun with a slightly crunchy top and a pillow-soft center, but it's also packed with protein, iron, B vitamins, and fiber. Go ahead and pile on the toppings; this bun can handle it!

ingredients

dry ingredients

1 cup brown rice flour
1 cup tapioca starch
½ cup arrowroot starch
½ cup millet flour
½ cup sorghum flour
1 tablespoon xanthan gum
1½ teaspoons sea salt

yeast mixture

1 cup warm water
(between 110° and 115°F)

3 tablespoons honey
2½ teaspoons dry active yeast

wet ingredients

3 flax eggs (page 10)
⅓ cup water
¼ cup olive oil
1 teaspoon apple cider vinegar
Sesame seeds for sprinkling

*The Hunzas, a people who reside in the Himalayan foothills and are known for their excellent health and longevity of life, use millet as a staple in their diet. Gluten-free ancient grains such as millet are typically less processed and contain more vitamins, minerals, and fiber than more common grains like corn, rice, and modern wheat. Millet can grow with little water compared to most other grains, making it well suited for drought-like conditions.

directions

1. Line two large baking sheet pans with parchment paper.
2. Combine the dry ingredients in a large bowl and mix well with a fork.
3. To activate the yeast, combine the honey and 1 cup warm water in a small mixing bowl and stir until the honey is mostly dissolved. Sprinkle in the yeast and stir. Set aside for 8–10 minutes until yeast proofs (foamy bubbles will appear).
4. While waiting, mix the eggs, ¼ cup water, oil, and vinegar together in a small bowl.
5. After yeast proofs, add egg mixture and yeast mixture to the large bowl and mix with a hand mixer until dough is smooth, about 1–2 minutes.
6. Before handling the sticky dough, place some warm water in a small bowl, and dip hands in the bowl. Pull about a ½ cup of dough from the mixture, create a disc shape, and place on the baking sheet. Continue to dip hands in the bowl while shaping 4 buns on each baking sheet. Slightly flatten the top of each bun and sprinkle with sesame seeds.
7. Allow buns to rise for about 1 hour in a warm, still area.
8. Preheat oven to 375 °F.
9. After the hour has passed, transfer buns to oven and bake for 20–25 minutes.
10. When buns start to turn golden brown, turn oven off.
11. Leave buns in and open the door for 15 minutes. (This will prevent buns from sinking.)
12. Remove buns, slice, and serve.

*
If you struggle with digestion issues and bloating, it could be worth a try to cut back on gluten even if you do not think you are gluten intolerant.

sean
SHERMAN

Sean Sherman has been on a mission to revitalize and bring awareness to Native American cuisine and Indigenous food systems.

In 2014, Sherman founded the catering and food education business The Sioux Chef to redefine North American cuisine and bring its often inaccessible culinary culture to as many communities as possible. In addition, he helped establish the nonprofit North American Traditional Indigenous Food Systems (NATIFS) focused on restoring Native foodways to address health and economic problems affecting Native communities. Sherman has won several awards and fellowships for his work including the 2018 James Beard Award for Best American Cookbook and the James Beard Leadership Award in 2019. In 2021, he and his business partner, Dana Thompson, opened Owamni, a restaurant featuring a decolonized menu of Indigenous foods and recipes.

Owamni was named the Minneapolis *Star Tribune's* 2021 Restaurant of the Year, and *Esquire* magazine featured Owamni as one of the "Best New Restaurants in America," also naming Sherman "Chef of the Year." In 2022, Owamni won a James Beard Award for Best New Restaurant.

Sean Sherman has been working in restaurants since he was 13 years old. However, he got inspired to explore the culture of Native American cuisine much later, during his time in San Panchos, Mexico. In this small town outside of Puerto Vallarta, Sherman came across an Indigenous group known as the Huicholes and was captivated by their art and mythology. He saw similarities between the way they lived and his own childhood growing up on a reservation in South Dakota. It was during this time that Sherman realized he had explored many different food cultures as an executive chef, but he didn't know much about his own ancestral food traditions.

Sherman returned to Minneapolis, where he'd been developing his culinary career, and began to build his business featuring Native American cuisine. To educate himself, he studied ethnobotany, spoke to elders and community members about how specific plants were used, and studied

Tȟuŋkášila (Grandfather),

I'm writing to let you know I have devoted my life and work to help all grandchildren of Indigenous peoples take the steps to reclaim our health, culture, and nutrition by revitalizing our Native foods. Pilamayeyelo (Thank you) for all of your strength and guidance and leaving us with stories of our ancestors. With everything our families had to endure, I hope we can make you proud knowing that we are carving a path for future Indigenous generations to celebrate our foods and for our non-Indigenous allies to help us protect the importance of revitalizing our Indigenous wisdoms.

Wopila Tanka Iciciyapelo...

sean

the clash between the U.S. government and various tribes across the country. He wanted to understand what kind of systems had been stripped away or damaged by colonialism.

Through his new restaurant, Owamni, Sherman is able to highlight the flavors of Native American cuisine and bring attention to the way Indigenous tribes have been cooking for hundreds of years. To open as many economic doors for Indigenous communities as possible, he sources many of his ingredients from nearby tribal farms, including chokecherries, elderberries, wild rice, and fish from the Red Lake Nation in Northern Minnesota.

When speaking about the food industry, Sherman notes how processed foods have taken over and dominated so many people's diets. He says that **Indigenous food traditions eliminate a lot of those processed foods like dairy, refined cane sugar, and wheat flour.** It "revitalizes the culture by looking at what was truly here before European influence came through." Owamni offers a decolonized menu of Indigenous foods and recipes. The dishes feature ingredients like wild rice from Minnesota and nixtamalized corn from Mexico. His dishes do not include colonial ingredients such as dairy, wheat, flour, sugar, beef, pork, or chicken. His goal is to bring back a taste of the past that was nearly destroyed by colonization.

Under NATIFS, Sherman also launched the Indigenous Food Lab. This Minneapolis-based lab is a professional Indigenous kitchen and training center that covers all aspects of food service. This includes research and development, Indigenous food identification, gathering, cultivation, and preparation, as well as every component of starting and running a successful culinary business based around Native traditions and Indigenous foods.

Beyond food service, Sherman explains that the Indigenous Food Lab, "is actually helping the tribes open up their own food businesses directly in their communities ... We help them to create food access in their communities."

The goal is to open up Indigenous Food Labs all across North America by picking large urban areas that can satellite out to tribal regions.

Through the success of Owamni and the development of the Indigenous Food Lab, Sherman has created two powerful avenues for promoting Indigenous food: a popular restaurant that creates awareness and appreciation for Indigenous dishes as well as a way to help tribes support and share foods within their own communities.

"I see a lot of oppression still alive today in Indigenous communities and a lot of issues surrounding nutrition and food because of oppression. I've seen a lot of health issues regarding this. It was a realization of looking backwards on how I grew up, and trying to design a path that can move forward."

SEAN SHERMAN'S
white tepary beans with nopales

pre-prep time: 8–10 hours to soak beans • *pre-cook time:* 2 hours to simmer beans

prep time: 10 minutes • *cook time:* 20 minutes to cook nopales (can be done while simmering beans)

assembly time: 10 minutes • *makes:* 5 cups or 6–8 servings

Tepary beans are amazing and so different from regular store-bought beans. They are incredibly drought resistant and have been grown for thousands of generations in the southwestern United States and northwestern Mexico. They take longer to cook than typical dried beans, but are highly nutritious and can be found in brown, white, or black. Enjoy this plant-based salad as a side or a main dish as you contemplate the Indigenous history of the land you currently occupy.

ingredients

1½ cups dried white tepary beans

2 fresh cactus paddles (nopales), cleaned

½ cup red onion, thinly sliced half moons

2 tablespoons sunflower oil

3 tablespoons pure agave syrup

1 tablespoon guajillo chile powder

1 teaspoon salt

1 tablespoon fresh oregano leaves

1 cup fresh dandelion leaves, cut into thin strips

directions

1. Place the white beans in a container with triple the amount of water to cover and let soak overnight.

2. Drain and rinse the beans the next day and place in a large pot and add 1:4 ratio of beans to water. Bring to a boil, then lower heat to a simmer and slow cook for 2 hours. When beans are tender, remove from heat, drain, rinse, and let cool.

3. To prepare the nopales, carefully scrape off any thorns with a sharp knife. Cut in half lengthwise and thinly slice. Place slices in cold water and bring to a simmer for 15–20 minutes, skimming mucilage until cactus stops secreting it. Remove from heat, strain, and rinse with cold water.

4. To assemble salad, toss cooled cooked beans, cooled cactus slices, and the rest of the ingredients until well incorporated.

Enjoy!

Tips

• Tepary beans can be purchased online from Ramona Farms.

• Cactus can be found in your local Mexican mercado.

• Make your own guajillo powder by toasting dry guajillo chilies, then pulverizing in a spice grinder or using a mortar and pestle.

sweet potato burger buns

pre-prep: 50–55 minutes to bake sweet potato • *prep time*: 20 minutes
rise time: 45 minutes • *bake time*: 18 minutes • *makes*: 10–12 buns

We know … we already have one gluten-free burger bun recipe, but we wanted to give you another option that is even more fun! The sweet potato not only adds a bit of color and texture, but also provides sweetness and added nutritional benefits. These buns are perfect for our Pulled "Pork" Sliders (page 94), but they pair great with any burger.

ingredients

1 large sweet potato

3½ cups gluten-free flour

2 teaspoons xanthan gum

1 teaspoon salt

1 tablespoon baking powder

3 cups rice milk

2 teaspoons honey or maple syrup

2 tablespoons dry active yeast

2 flax eggs *(page 10)*

2 tablespoons coconut sugar

¼ cup dairy-free butter, melted

* *Surprisingly, coconut sugar comes from the sap of the coconut palm tree and not the actual coconut. It's a good alternative to cane sugar, as it retains many nutrients found in the coconut palm, such as iron, zinc, calcium, and potassium. We still suggest that you use coconut sugar in moderation, since at the end of the day it is still sugar and is high in fructose.*

directions

1. Preheat oven to 400°F. Wrap sweet potato loosely with foil and bake for 50–55 minutes, until potato is very soft. Let potato cool, set aside. Reduce oven temperature to 375°F.
2. In a large bowl, combine flour, xanthan gum, salt, and baking powder. Mix with a fork and set aside.
3. Pour rice milk and honey in small pot and warm until 115°F. Pour mixture in small mixing bowl and add yeast. Mix to combine and let sit for 8 minutes, until the mixture is foamy.
4. Line 2 baking sheets with parchment paper and set aside.
5. Mash the sweet potato (skin included) until smooth, and add to large bowl of flour mixture, along with flax eggs, rice milk mixture, coconut sugar, and melted butter. Mix with hand mixer for 2 minutes.
6. Make a small bowl of warm water for hands. Dip hands in water and form about ½ cup of dough into a ball shape and place on baking sheet. Flatten out top. Make 5–6 flattened-out ball shapes on each baking sheet, making 10–12 buns. Place a foil cover on each sheet and allow buns to rise in a warm, draft-free place for 45 minutes.
7. Bake buns for 17–18 minutes, until golden brown. Turn oven off.
8. Leave buns in and open the door for 5–10 minutes before serving.
9. Remove buns, slice, and serve.

Sweet potatoes are slightly higher in fiber, vitamin B6, and vitamin C when compared to white potatoes. White potatoes are higher in potassium, but sweet potatoes contain much more vitamin A.

cassava flour tortillas

prep time: 10 minutes • cook time: 10 minutes • makes: 5 tortillas

Tortillas are a great staple to have in your kitchen and can be used for a variety of recipes, such as breakfast tacos, Sweet Potato Enchiladas (page 112), or Sweet Potato Chickpea BBQ Wraps (page 99). Plus, cassava flour serves as a wonderful wheat replacement and makes these tortillas gluten-free.

ingredients

1 cup cassava flour

½ cup coconut milk

¼ cup olive oil

¼ cup water

1 teaspoon sea salt

1 teaspoon garlic powder

½ teaspoon pepper

extra oil for frying

directions

1. Combine all ingredients in a medium bowl and mix together until well combined.
2. Divide the dough into 5 equal parts and shape into balls. On a piece of parchment paper dusted with cassava flour, pat each shape into a thin tortilla. Using your hands to shape them works well, but dust them first with cassava flour.
3. Heat fry pan to medium heat. Drizzle pan with oil of your choice.
4. Cook tortillas for 1–3 minutes on each side, until tortilla begins to brown. Let cool and serve.

*Cassava flour and tapioca flour are both made from the cassava root. While tapioca flour is extracted from the cassava root through a process of washing and pulping, cassava flour is the whole root, simply peeled, dried, and ground, giving it more dietary fiber than tapioca flour.

sweet potato biscuits

pre-prep: 50 minutes to bake sweet potato • prep time: 15 minutes
baking time: 24 minutes • makes: 8–12 biscuits

We used to love traditional biscuits, but they're loaded with butter, milk, and white flour. Our plant-based version with sweet potatoes and coconut milk gives this recipe great flavor and nutrients, and we've also made it gluten-free with a variety of flours. Our biscuits have a soft and tender texture with the subtly sweet and buttery taste of traditional-style biscuits, without the heavy feeling afterwards. Have one with jam at breakfast, pair it with soup, or dip it in our Cashew Mushroom Gravy (page 185).

ingredients

1 medium sweet potato with skin

1½ cups gluten-free flour

¼ cup sorghum flour

¼ cup coconut flour

1 tablespoon baking powder

1 teaspoon guar gum

1 teaspoon cinnamon

½ teaspoon sea salt

1 cup nut milk (*we use coconut*)

½ cup maple syrup

½ cup palm shortening
We recommend Nutiva Organic Shortening,
a mix of coconut and palm oil that is sustainably harvested.

directions

1. Preheat oven to 425°F. Pierce entire sweet potato with a fork and bake for 45–50 minutes.
2. Line baking sheet with parchment paper. In a large mixing bowl, combine all of the flours with baking powder, guar gum, cinnamon, and sea salt.
3. When sweet potato has cooled, leave skin on and mash lightly with a fork (a few lumps are fine) and place in a medium bowl. Stir in milk and maple syrup.
4. Blend the palm shortening into the dry ingredients, and mix until it is blended well. Add the sweet potato and milk mixture and stir together until blended.
5. Break off pieces of the mixture and form 8–11 balls. Place each ball onto the parchment paper and slightly push down until the top is a little flatter.
6. When all of the mixture is formed into the round shapes, bake for 20–22 minutes. Cool and serve.

Sorghum is an ancient crop of African origin and is especially important in arid and semi-arid regions, as sorghum is both drought- and heat-tolerant. It's the fifth most important cereal crop in the world.

roasted bell peppers

prep time: 5 minutes • *baking time:* 30 minutes • *serves:* 5

This side dish is the perfect complement to many meals, such as the Chickpea "Crab" Cakes (page 60), Sweet Potato Gnocchi (page 111), or the Black Bean Beet Burger (page 127). It's fun to try this recipe with different colored bell peppers too, as yellow bell peppers have the highest vitamin C content, while red bell peppers are high in beta-carotene. Eat the rainbow!

ingredients

4 bell peppers, assorted colors
2 tablespoons olive oil

directions

1. Preheat oven to 400°F. Place whole bell peppers on a baking sheet and lightly coat each pepper with oil.
2. Lay peppers on side, bake for 15 minutes.
3. Using tongs, turn peppers and bake another 12–15 minutes.
4. Slice pepper vertically to open and gently pull the stem from the top. The stem and a clump of seeds should loosen easily.
5. Strip off charred skin. These peppers can now be sliced and served.

* Since *bell peppers* come from *blooming plants and have seeds, they are actually a fruit, not a vegetable!*

garlic cauliflower / potato mash

pre-prep: 40 minutes to bake garlic, 1 hour to bake potatoes, 8-10 minutes to steam cauliflower

prep time: 10 minutes • serves: 4

Nothing says comfort food like garlic mashed potatoes. Traditionally, this dish is loaded with milk, butter, and sour cream, giving it the richness that tastes so good but tends to leave you with a heavy feeling. Our version leaves out the dairy but maintains the savory goodness and also swaps out half the potato for cauliflower, leaving you with a lighter feeling than the classic recipe.

ingredients

2 garlic bulbs

½ tablespoon olive oil

5 medium potatoes or 7 small potatoes,
   Yukon Gold or red

2 cups cauliflower florets

3 tablespoons vegan butter

¼ cup plant-based sour cream
   (To make our Cauli/Cashew Sour
   Cream, see page 186)

¼ teaspoon salt

¼ teaspoon pepper

topping ideas

Cashew Mushroom Gravy (page 185)

directions

1. Preheat oven to 400°F. Slice off tops of garlic bulbs and remove a few of the outer layers of skin. Drizzle olive oil on top of bulbs. Wrap garlic in parchment paper (or foil) and place directly on oven racks for 30–40 minutes.

2. Using a medium saucepan and steamer basket, fill bottom of pan with water and steam potatoes covered for 15–30 minutes, depending on size of potatoes (see page 12 for steamer directions).

3. Steam cauliflower florets in a steamer basket for 6–8 minutes until soft (see page 12 for steamer directions).

4. In a small pot, melt butter on low, set aside.

5. Place potatoes in a large bowl and mash with a potato masher or a large fork until potatoes are mashed but still have texture. Add cooked cauliflower and continue to mash. Stir in sour cream, butter, salt, and pepper.

6. When garlic has cooled, squeeze the garlic out of its skins and mash in a small bowl. Blend in with potato mixture.

7. Serve immediately or refrigerate. You can also freeze mash to save for later.

bryant TERRY

Bryant Terry is a nationally acclaimed chef, educator, and author renowned for his activism to create a healthy, just, and sustainable food system.

Terry is editor-in-chief of 4 Color Books, an imprint of Penguin Random House's Ten Speed Press. He's also the founder, co-principal, and innovation director of the creative studio Zenmi. Since 2015, Terry has been the Chef-in-Residence at the Museum of the African Diaspora (MoAD) where he creates public programming at the intersection of food, farming, health, activism, art, culture, and the African Diaspora. Terry received the 2015 James Beard Foundation Leadership award for his food justice work. He is the author of six cookbooks including *Afro-Vegan*, *Vegetable Kingdom*, and his latest, *Black Food*. In 2021, *Black Food* was lauded as one of the most important cookbooks of the year by the *New York Times*, *NPR*, *SF Chronicle*, and others. *San Francisco Magazine* included Terry among 11 Smartest People in the Bay Area Food Scene, and *Fast Company* named him one of 9 People Who Are Changing the Future of Food.

Terry uses cooking as a tool to illuminate the intersection of poverty, structural racism, and food insecurity. He grew up in Memphis, Tennessee, with agrarian roots in the rural south, which he believes to be one of the greatest privileges of his life. His family and neighbors had gardens and ate off the land. Terry's family also had farmland in different parts of the South. Eating locally, seasonally, and organically was his family's ethos. It wasn't something that was celebrated; it simply *was*. In high school, Terry began to eat unhealthy foods, until a friend made him listen to a song that changed his life: "Beef," a hip-hop song about animal cruelty in factory farms. That is when Terry began to connect the dots around our dysfunctional food system. When he began attending food conferences, he realized that the most impacted people were not present. So, he founded b-healthy! (Build Healthy Eating and Lifestyles to Help Youth!), a multi-year initiative in New York City that educated youth about healthier food systems. One of his mentees went on to start her own educational program to empower youth.

The Black Panthers also formed Terry's outlook as he believes that grassroots movements can shift public policy at both the local and federal levels. Thus, Terry focuses on empowering the communities that are impacted most intensely by our food system. He is passionate about putting the *culture* back into agriculture.

Although work and passion led him to become a national presence, Terry still experienced some guilt for not being rooted in Memphis, the place where he grew up. Van Jones, a mentor of his, helped Terry understand that sometimes, being away and having a national presence is needed. Fighting systemic problems requires people to play their part on all levels. This includes everyone from grassroots organizers to philanthropic funders, elected officials to public speakers, journalists to chefs.

His book *Vegetable Kingdom* was published in February 2020, right before the pandemic hit. The book could have been overlooked during such a hectic time, but instead, people turned to cooking. *Vegetable Kingdom* helped folks eat in a healthier and more mindful way when they were forced to stay home. The book went on to be a commercial and critical success, winning an NAACP Image Award, and it was named one of the best cookbooks of the year by outlets like *The New Yorker*, the *Washington Post*, *Vogue*, and *Food & Wine*.

In May of 2021, West Coast publisher Ten Speed Press announced that Terry would lead an imprint called 4 Color Books, which brings two to three titles, written by diverse authors, to market each year. In an October 2021 interview with the *New York Times*, Terry says that this new role "is formalizing what [he's] been doing for years, which is mentoring younger writers, helping people of color who are thinking of writing a book." He sees his imprint as, "a pipeline for Black food creatives and people of color in publishing."

> Terry uses cooking as a tool to illuminate the intersection of poverty, structural racism, and food insecurity.

Terry has a specific philosophy about how best to engage people. He says, "I think that starting with politics and heady intellectual ideas can often alienate people. It's not as effective as starting with things that resonate with everybody, like food, like culture. Those are the things that I've always used to open the conversation about changing our food system and getting people to think about the role that they play in shifting public policies around food systems."

The first title from 4 Color Books, Terry's own *Black Food*, released in 2021, showcases the breadth of Black culture around the world. The book is much more than a cookbook as it features an array of Black voices through recipes, stories, artwork, and more. *Black Food* became the most critically acclaimed American cookbook published that year.

Terry's previous cookbooks, including *Vegan Soul Kitchen*, *Afro-Vegan*, and *Vegetable Kingdom* have focused on vegetarian and vegan African American cooking. These titles created a contrast to public perception about Black cooking and its history. "Enslaved Africans' relationship to food and cooking was shaped by a number of factors," Terry told the *New York Times*, "including geographic location and the financial status and disposition of plantation owners." Because cooking during the era of slavery was different in the Carolinas, the Caribbean, and Louisiana, to call it "slave food," he said, is "a misinformed, reductive, and racist way to frame Black cuisine."

Like many businesses, the publishing industry was profoundly impacted by the racial reckoning of 2020, which is why there is no better time for Terry to come out with a publishing imprint for BIPOC authors. His work as a chef has evolved in so many expansive ways, and we're excited to follow 4 Color Books and to see Terry's future projects in the world of publishing.

"What has been my guiding philosophy since I started doing this work is this idea of starting with the visceral, to ignite the cerebral, and ending at the political."

Dear Paw Paw,

I often tell people that everything I've been teaching over the past two decades I learned from you as a child, and I uplift the cooking and eating practices that were just a part of our family culture. You know, I don't recall you ever mentioning the kind of buzzwords that we often hear when talking about building a more healthful, just, and sustainable food system, but I am clear that we ate food that was as local as our backyard gardens, mostly in season (save for what we canned, pickled, and preserved), and I remember us literally harvesting food right before our meals. While I didn't always appreciate the times spent in your home garden planting seeds, pulling weeds, watering dark leafy greens, shucking corn, shelling peas, and the like, I look back at those times with so much fondness. I feel so blessed to have experientially learned about the seed-to-table cycle from you. I also learned the value and power of growing one's own food, and I developed an ethos of sharing from seeing you provide family, friends, and church members with surplus from your garden. I know that many Black folks from the South have similar stories of growing up in families with agrarian roots in rural areas, and I like to think I am helping people re-member, piece back these histories that our industrialized food system has made us forget. I hope that I make you proud with all I've accomplished over the past two decades. I stand on your shoulders.

Love,

bryant

BRYANT TERRY'S

hibiscus-ginger shrub

prep time for 1st day: 10 minutes, allow to sit for 24 hours
prep time for 2nd day: 10 minutes, allow to sit another 24 hours
cooking time: 5 minutes • **makes:** 5 servings

ingredients

hibiscus-ginger vinegar

2 cups dried hibiscus
2 heaping tablespoons grated fresh ginger
6 whole allspice berries
2 cups apple cider vinegar
1 cup raw organic cane sugar

shrub

crushed ice
soda water
limes for garnishing

directions

make the vinegar

1. Put 1 cup of the hibiscus, ½ of the grated ginger, and three of the allspice berries into a small stainless steel saucepan. Pour the vinegar over the solids, cover, and store in a cool place for 24 hours.
2. Strain the vinegar and set it aside. Compost the solids. Put the remaining hibiscus, ginger, and allspice berries into the receptacle just used, pour over the vinegar, cover, and store in a cool place for another 24 hours.
3. Strain the vinegar and set it aside. Compost the hibiscus just used. Combine the vinegar and sugar in the saucepan. Over high heat, bring to a boil. Reduce to medium-high and simmer until hot to touch and the sugar has dissolved, 3–5 minutes. Set aside and cool. Transfer to a sterilized bottle and refrigerate until ready to use.

make the shrub

1. Fill a 10–12 ounce glass with crushed ice. Pour in ¼ cup Hibiscus-Ginger Vinegar and top off with an equal amount of soda water.
2. Garnish with lime and serve immediately.

why ginger?

Ginger is one of the healthiest spices on the planet. It has been used in various forms of traditional medicine for thousands of years, and the main bioactive compound, gingerol, has powerful anti-inflammatory and antioxidant effects.

miso-glazed eggplant

prep time: 15 minutes • **soak time:** 1 hour • **cook time:** 20 minutes • **serves:** 3-4

We'd never been so excited about a vegetable side dish until we made this recipe. The combination of the wine and miso make it taste like ... eggplant french toast! Okay, we know that sounds bizarre, but it's just so delicious. In addition, eggplants have high levels of antioxidants and fiber. The recipe calls for the eggplant slices to soak for 1 hour, but it's also fine to soak them in the morning and let them absorb the flavors all day, then fry or bake them when you're ready to serve. Get ready to wow your guests with this one!

ingredients

1 large globe or Italian eggplant

¾ cup sake or white wine

¼ cup rice wine (*mirin*)

¼ cup water

6 tablespoons white miso

1 teaspoon honey

1 tablespoon coconut oil, melted

coconut oil for frying

directions

1. Lay eggplant on its side and slice into ¼" discs. Set aside in a shallow bowl.
2. Mix the rest of the ingredients together and pour over eggplant. Marinate in sauce for 1 hour.
3. Coat skillet with coconut oil. Fry each eggplant slice on low for 5 minutes on each side, or until golden brown, or roast in the oven on a baking sheet at 375°F for 15–20 minutes. Serve immediately. The remaining sauce can be refrigerated and saved for up to a week.

*** Miso, a fermented paste made from soybeans** and a koji starter, is believed to have originated in China and was later introduced to Japan by Buddhist monks. The fermentation process involved in miso production includes beneficial bacteria known as probiotics, which can promote digestion and gut health.*

sweet potato harissa soup

pre-prep: 55 minutes to bake sweet potato • *prep time*: 15 minutes
simmer time: 10 minutes • *serves*: 2-4

Sweet potatoes are in season from fall to early winter, which is the perfect time to make this soup. Even with just a few ingredients, when it comes to flavor it's a dish that's hard to beat. The creamy texture from the sweet potato, along with the aromatic spices and unique harissa taste, offers the perfect combination of sweet and spicy.

ingredients

1 large sweet potato
1 tablespoon coconut oil
1 onion, sliced
1 teaspoon ground cumin
2 cups vegetable broth

1 cup coconut milk
Salt and pepper to taste
2 teaspoons harissa paste
pinch of red pepper flakes

serving ideas

Coconut Bacon (*page 171*) • Cornbread (*page 132*)

directions

1. Preheat oven to 400° F. On a baking sheet, pierce sweet potato all over with a fork. Bake for 50–55 minutes, until tender. When cooled, cut into medium chunks.
2. Heat oil in a large pan over medium heat. Add in onion slices and cook, stirring occasionally for about 5 minutes. Add cumin and cook another minute.
3. Add onion mixture to blender or food processor, along with cooked sweet potato chunks, vegetable broth, and coconut milk. Add a pinch of salt and pepper and blend well.
4. When mixture is completely smooth, add to medium saucepan and simmer for 10 minutes or until it reaches desired temperature.
5. When ready to serve, transfer to individual bowls and add a swirl of harissa paste and red pepper flakes.

Cumin is one of the most commonly used spices in the world. In addition to its value as a seasoning, cumin is used in supplements and herbal teas.

alice
WATERS

Alice Waters is the visionary chef, activist, and restaurant owner of Chez Panisse restaurant in Berkeley, California.

An author and a strong advocate for regenerative organic agriculture, she has played a key role in advancing projects related to the cause. Waters created the Edible Schoolyard Project, an organization that advocates for a free regenerative organic school lunch for all children and a comprehensive food curriculum in every public school. It also opens opportunities to teach the values of nourishment, stewardship, and community. Most recently, Waters helped to open Lulu, a new restaurant in the Hammer Museum in Los Angeles. With over 16 books and many honors such as the National Humanities Medal, Waters has proven that food is a powerful way to create positive change and promote social justice.

In 1994 Alice Walters founded the Edible Schoolyard at Berkeley's Martin Luther King Jr. Middle School, which integrates **organic gardening and cooking into academic classes** and school life. It includes a kitchen classroom where students cook meals with the produce they grow. In the classroom, they may learn about the history of the Middle East while eating hummus they've made themselves, or count seeds in the garden for a math class, or perhaps learn chemistry while making fresh ricotta.

This is just one of many ideas Waters has enacted. For over four decades, she has been pushing for the importance of regenerative agriculture and local sustainable farming, inspired by a powerful combination of culture and community. After spending time in Paris and taking part in the free speech rally in Berkeley back in 1964, Waters was captivated by a bigger picture that, she says, "woke up [her] senses." Her time in Europe had given her a new understanding of food. She wanted a farmers' market where she could buy her food seasonally. With the added influence of Elizabeth David's cookbooks, which she claims are responsible for the way she looks at food, Waters was set on a path that would dramatically change the movement for local, organic farming for the better.

When she first opened Chez Panisse in 1971, the idea of "organic" was not as popular as it is now. At first, Waters simply wanted people to taste things they had never tasted before. She felt that if the food was good, people would come and the restaurant was ultimately successful as a result. Now, her restaurant is famous for its organic, locally grown ingredients and strong support for the farm-to-table movement. According to *The New York Times*, Waters hasn't shopped at a conventional grocery store in 25 years!

Dear Maria Montessori,

Your teachings were instrumental in making Chez Panisse what it is today, and were the true inspiration to starting the Edible Schoolyard Project in Berkeley, CA.

Your philosophies of our senses being the pathways to our mind, that the best way to learn is simply by doing, and beauty as a language of care, are my core beliefs.

Thank you for awakening my senses.

With gratitude,

alice

ALICE WATERS'
minestrone soup

pre-prep: Soak the beans overnight in a large pot, covered by several inches of water.
The next day, simmer the beans for 2 hours or until tender. Drain and set aside, reserving the cooking water.

prep time: 15 minutes • *cook time*: 45 minutes • *makes*: 8 servings

ingredients

1 cup dried cannellini or borlotti beans
   This will yield 2½–3 cups of cooked beans.

¼ cup olive oil

1 large onion, finely chopped

2 carrots, peeled and finely chopped

4 garlic cloves, coarsely chopped

5 thyme sprigs

1 bay leaf

2 teaspoons salt

3 cups water

1 small leek, diced

½ pound green beans, cut into 1-inch lengths

2 medium zucchini, cut into small dice

2 medium tomatoes, peeled, seeded, and chopped

1 cup bean cooking liquid

2 cups spinach leaves, coarsely chopped *(about 1 pound)*

for garnish

extra-virgin olive oil • parmesan cheese

directions

1. Heat olive oil in a heavy-bottomed pan over medium heat. Add onion and carrots and cook 15 minutes or until tender.

2. Add garlic, thyme, bay leaf, and salt. Cook for 5 minutes longer. Add water and bring to a boil.

3. When boiling, add leek and green beans. Cook for 5 minutes, then add zucchini and tomatoes. Cook for 15 minutes, taste for salt, and adjust as necessary.

4. Add the cooked beans along with bean cooking liquid and spinach leaves. Cook for 5 minutes. If the soup is too thick, add more bean cooking liquid. Remove the bay leaf.

5. Serve in bowls, each one garnished with 2 teaspoons extra virgin olive oil and 1 tablespoon or more grated Parmesan cheese. Pesto is another excellent garnish for the soup.

To make it plant-based:
• vegan parmesan cheese can be substituted for parmesan cheese

* Cut all the vegetables into *bite-size pieces* so that each spoonful will have a variety of tastes and textures.

According to The New York Times, *Waters hasn't shopped at a conventional grocery store in 25 years!*

Waters is currently working on the initiative for free, regenerative organic lunches in the state of California and trying to bring attention to the matter any way she can. Whether she's communicating with former U.S. Senator Barbara Boxer or finding people who are running for office to sign her pledges, she feels strongly about getting her message out there. Her Edible Schoolyard Project has reached over 5,800 schools around the world and she hopes to grow that number even further so that every student can be guaranteed a free and nutritious meal.

Guided by a strong sense of purpose, she uses the power of food to do good in the world. "I really believe in a free, sustainable school lunch for every child K–12. I want schools to support the farmers, the ranchers, the fishermen, the people who are taking care of the land for future generations." Her vision is all encompassing and she envisions a community that feeds children "the most nourishing food and really supports the farmers who are doing the right thing." She advocates for a system that teaches "the values of stewardship, and community, and nourishment as part of the curriculum in every school in this country."

The culmination of Alice's lifetime of work is the founding of the Alice Waters Institute for Regenerative Agriculture and Edible Education (AWI) in partnership with the University of California. Alice and AWI seek to profoundly change the way food is conceived, sourced, and served in our public schools to better nourish our children and to address our climate crisis through school-supported regenerative agriculture.

sweet potato hash browns

*prep time: 30 minutes • **bake time**: 40 minutes • serves: 10*

As a classic American breakfast side dish, hash browns totally rock. Aside from the recollection of a familiar diner, though, they might also make you think of a dish bathed in oil with little-to-no nutritional benefit. It doesn't have to be this way. We wanted to create a version that adds sweet potatoes into the mix and not only provides that home-style comfort, but also includes a variety of veggies and nutrients that keep you full and satisfied.

ingredients

1 large sweet potato

2 Yukon Gold potatoes

2 medium carrots, peeled and grated

1 onion, finely chopped

3 garlic cloves, minced

handful of fresh parsley, chopped

4 ounces vegan cheese, grated

1 flax egg *(page 10)*

1 tablespoon chickpea flour

½ teaspoon salt

½ teaspoon pepper

1–2 tablespoons coconut oil

topping ideas

applesauce • Cauli/Cashew Sour Cream *(page 186)* • Vegan Fried Egg *(page 177)*

directions

1. Preheat oven to 375°F. Line 2 baking sheets with parchment paper.
2. Using a grater, grate potatoes and place in a mesh bag or towel and squeeze out as much moisture as possible.
3. Add potato gratings to mixing bowl, along with grated carrots, onion, garlic, parsley, cheese, flax egg, chickpea flour, salt, and pepper. Mix all ingredients.
4. Take a small handful of mixture and form a rectangle. Make 5 rectangles on each baking sheet.
5. Bake for 40 minutes until mixture becomes golden brown.
6. Pour about ½ to 1 tablespoon coconut oil in a large frying pan and heat to medium/high. Fry each hash brown for 2 minutes on each side before serving. After 5 are done, pour ½–1 tablespoon coconut oil in the pan and fry the next 5.

lentil walnut meatless meat

pre-prep: 20 minutes to cook lentils (page 11), 10 minutes to soak walnuts
prep time: 10 minutes • *makes*: 2 cups

Lentils are nutritional powerhouses, loaded with protein, potassium, folate, iron, and fiber. In addition, the lentil crop has an extremely low carbon footprint and is a "nitrogen fixer" that converts atmospheric nitrogen into useful ammonia or nitrates, improving soil fertility and reducing the need for energy-intensive fertilizers. This dish is a fantastic meat substitute that we call for in our Sloppy Joes (page 115), or it can be added to our Marinara Sauce (page 178) to serve over pasta!

ingredients

1 cup walnuts

1 teaspoon olive oil

3 cloves garlic, minced

2 teaspoons paprika

½ tablespoon liquid aminos

¼ teaspoon salt

¼ teaspoon pepper

2 tablespoons olive oil

2 tablespoons water

½ tablespoon sriracha sauce

2 tablespoons maple syrup

¼ teaspoon sesame seeds

1½ cups lentils, green, cooked
Red are a little soft, but can still work

directions

1. Soak walnuts in water for 10 minutes. Drain.

2. Add 1 teaspoon olive oil to a small pan, sauté garlic for 3 minutes.

3. In a medium bowl, add sautéed garlic, paprika, and all ingredients other than the lentils and walnuts. Mix to combine.

4. In a food processor, add lentils, walnuts, and the mixed seasonings. Pulse 10–12 times, until lentils are slightly blended, but not mushy. Heat mixture on low for 5 minutes and serve.

* *We recommend lentils from Timeless Natural Food, founded by David Oien, who is featured in the book* Lentil Underground *written by Groundbaker Dr. Liz Carlisle (page 90).*

sweet or savory coconut bacon

prep time: 10 minutes • **bake time**: 12–14 minutes • **cool time**: 30 minutes

Coconut Bacon is a great topping to have on hand for your meals. We have two choices here, either sweet or savory depending on what you're feeling. We suggest adding this bacon to our Mac and Cheese (page 74) and our Southwest Salad (page 96), but it's also delicious with a BLT, soup, salad, or pasta!

ingredients for sweet bacon

2 cups large unsweetened coconut flakes

2 tablespoons coconut aminos

⅔ teaspoon reduced-sodium tamari

1½ tablespoons liquid smoke

1 tablespoon maple syrup

directions

1. Preheat oven to 350°F. Line a large baking sheet with parchment paper.
2. Spread coconut flakes evenly onto baking sheet. Mix remaining ingredients in a small bowl and drizzle over coconut flakes. Toss flakes to make sure they are evenly coated. Then spread out evenly to bake.
3. Bake for 12–14 minutes, flipping halfway until flakes are mostly dry and turning golden on the edges.
4. Turn oven off, but leave bacon in oven for 20 minutes to crisp with oven door slightly open.
5. Remove bacon from oven and let flakes cool for 10 minutes.
6. Store leftovers in the refrigerator or freezer.

ingredients for savory bacon

2 cups large unsweetened coconut flakes

juice from 1 lime

½ teaspoon cayenne pepper

½ teaspoon salt

directions

1. Preheat oven to 350°F. Follow directions for the Sweet Coconut Bacon.

 With about 60% less sodium, *coconut aminos* are a great replacement for soy sauce. They're also soy-free and gluten-free for those with food allergies.

DR. *daphne* MILLER

Daphne Miller, M.D., is a practicing family physician, clinical professor at the University of California, San Francisco, and integrative and community medicine curriculum director at the Lifelong Medical Family Medicine Residency Program in Richmond, California.

Miller also directs the Growing Health Collaborative, a program based at the University of California, Berkeley, which engages health professionals in transforming the food system from the soil up. She is the author of two best-selling books: *The Jungle Effect* and *Farmacology*. Her writing and teaching explore the frontier between medicine and the natural world.

Miller started her edible journey as a child of Peace Corps parents. Growing up in North Africa and the Middle East, she saw how food served as a universal language to help her navigate other cultures and make new friends.

In medical school, she was dismayed to see how little attention was placed on food and diet. When food was discussed, it was presented as a collection of nutrients removed from its broader context of culture and gastronomy. "I realized humans eat *food*," she says, "not just micro and macro nutrients."

As a family physician in San Francisco, she frequently saw patients with heart disease, diabetes, or other chronic ailments who were the first generation in their family to have these health problems. They often reported that their parents and grandparents had grown up in other parts of the world. Curious about this paradox, Miller began to research her patients' ancestral diets, and this culminated in *The Jungle Effect,* a book about the healthiest diets from around the world.

Dr. Miller wants to help people discover their "inner doctor," the skills they need to manage their health. Miller believes that reconnecting to how our food is grown can be powerful medicine. Her book *Farmacology, Total Health from the Ground Up* helped spark a movement to explore how healthy farms, healthy land, and healthy soil can protect us while protecting the planet.

"For a long time," she says, "I was a food-is-medicine doctor, talking a lot about food and recipes and the connection to health. Somewhere along the way, it dawned on me that I was only telling half the story, and that so much that has to do with our health isn't about the broccoli; it's about the varietal of that broccoli, and the history of

> ### *"I like to work with folks to help them retrain their taste buds."*

that broccoli. Where was it grown? How was it grown? What water and chemicals or lack thereof were used in growing it? What were the companion crops in the field? What is the story of the farm workers that worked to produce that broccoli? What are the national policies that keep that broccoli from being grown in the first place or make it much more likely that a less nutritious crop would be grown in that acre of land? I realized that all of these questions were much more relevant to our health than simply the nutrients in a stalk of broccoli."

Asked about dietary advice she gives her patients, Miller said, "I like to work with folks to help them retrain their taste buds. Our food landscape is so flooded with sugar and salt that things that are bitter, astringent, sour, and savory just taste wrong. But the truth is that these are the most interesting foods, and the healthiest for us." (She plucks off a leaf from a dandelion growing near her feet.) "This is what we should be eating!"

Dear Nature,

Thank you for sharing your instruction book with me. You have helped me become a better doctor and a better gardener. From you I have learned to respect the natural patterns of things, support existing synergies, and promote diversity.

So much love,

daphne

DR. DAPHNE MILLER'S

jushi rice

prep time: 10 minutes • ***cook time:*** 55 minutes • ***makes:*** 6 servings

This dish is a favorite among Okinawan children. While the traditional recipe calls for purple Okinawan potatoes, it can also be made with other types of potatoes or root vegetables.

ingredients

2 cups short-grain brown rice*

1 Okinawan sweet potato (*purple potato*),
   cut into ¼-inch cubes

2 tablespoons soy sauce, low sodium

¼ cup dried shiitake mushrooms,
   stems removed, chopped, or
   1 cup fresh shiitake mushrooms

2½ cups water

directions

1. Add rice, potato cubes, soy sauce, and shiitake mushrooms to a pot of water over medium/high heat. Cover and bring to a boil.

2. When mixture starts to boil, turn heat down and simmer for 50 minutes.

3. Remove lid and turn up heat for 10 seconds. Remove from heat, and serve immediately.

* *This recipe works well with other types of rice, such as black rice (forbidden rice), brown rice, basmati, or white rice.*

Recipe reprinted with permission from The Jungle Effect

eggplant pepperoni

prep time: 10 minutes • **soak time:** 20 minutes • **bake time:** 10 minutes • **makes:** 50 slices

M eat eaters and vegetarians alike will love making this savory and satisfying topping for the Pepperoni Pizza Pinwheels (page 84) or using it as a pizza topping! We recommend Chinese eggplant, which is the perfect size to replicate pepperoni slices.

ingredients

1 medium Chinese eggplant

2 tablespoons olive oil + oil for baking or frying

1 tablespoon liquid aminos

¼ teaspoon fennel, ground

½ teaspoon paprika

1 teaspoon nutritional yeast

pinch of white pepper

½ teaspoon garlic powder

½ teaspoon onion powder

¼ teaspoon anise, ground

directions

1. Preheat oven to 425°F. Lightly grease baking sheet and set aside unless you prefer to fry the eggplant. Cut eggplant into thin slices, about 45-50 pieces. Set aside.

2. In a small bowl, add all of the other ingredients and mix well. There will only be a small amount of marinade, but that is all you need, since the flavor is very strong. For best results, mix with hands.

3. Place eggplant slices in bowl and marinate for 20 minutes.

4. Once eggplants are finished marinating, choose to either bake or fry slices.

5. For baking: Lay eggplant slices on baking sheet, bake for 10 minutes until golden brown. Cool before serving.

6. For frying: Lightly spray large pan with oil and cook on high heat for 5–8 minutes, turning halfway through.

7. If using this recipe for Pepperoni Pizza Pinwheels, see page 84.

8. If using this recipe for pizza, bake slices for an additional 5 minutes or fry for an additional 2–3 minutes before placing on unbaked pizza.

vegan fried egg

prep time: 10 minutes • *cook time*: 10 minutes • *makes*: 8–10 eggs

To make a Vegan Fried Egg for Matt's Loco Moco recipe (page 80), we wanted to combine healthy ingredients that would look and taste like the real thing. We tried many combinations and found this one to be similar not only in appearance, but also taste! Thanks to black salt (kala namak), which we use in our Eggless Salad Sandwich (page 67), we feel we've created a great replica. These eggs can be added to Blueberry Banana Pancakes (page 22), Gluten-Free Bagels (page 40), or Sweet Potato Hash Browns (page 169).

ingredients

egg whites

12 ounces firm tofu, drained

½ cup white rice flour

2 tablespoons cornstarch

1 tablespoon olive oil

½ teaspoon black salt (*kala namak*)

½ teaspoon regular salt

½ teaspoon white vinegar

¾ cup water

egg yolk

½ cup of the egg-white mixture

½ teaspoon black salt (*kala namak*)

1 teaspoon yellow mustard

4 tablespoons nutritional yeast

½ teaspoon turmeric

¼ teaspoon paprika

avocado oil or olive oil spray for frying

directions

1. Drain tofu and add to a blender, along with all other egg-white ingredients. Blend until smooth. Transfer to a medium bowl.

2. For the yolk, add ½ cup of egg-white mixture to a small bowl and whisk together with the rest of the egg-yolk ingredients.

3. Measure out 3 tablespoons of egg white and put in a small bowl. Spray large frying pan with a good coating of oil and heat on medium/high. Make sure pan is well heated. When hot, add the measured-out egg-white mixture to pan. Make a small indentation in the middle of the mixture with a spoon. Scoop 1 tablespoon of egg yolk into the center of the egg white. Cook on one side for 2–3 minutes, then carefully flip the egg over for another minute. Lightly flatten the cooked side of the egg with a spatula.

4. Continue spraying pan and frying eggs until batter is done. Serve immediately, or store in refrigerator for 2–3 days.

marinara sauce

prep time: 15 minutes • *cook time*: 15 minutes • *makes*: 4 ½ cups • *serves*: 4–5

This classic sauce is easy to make and a game changer in many ways. Once we began making our own fresh sauce, we could easily taste the difference compared to store-bought marinara. This sauce is made with a blend of herbs and seasonings that perfectly complement the slow-simmered, rich tomato marinara flavors. Use this sauce in our Sloppy Joes (page 115), Pepperoni Pizza Pinwheels (page 84), or Sweet Potato Gnocchi (page 111).

ingredients

1 jar (18-ounces) crushed tomatoes
  We recommend Jovial Organic Crushed Tomatoes.

3–4 tomatoes, sliced

½ cup low-sodium vegetable broth

1 cup onion, chopped

3 garlic cloves, minced

2 tablespoons olive oil

¼ teaspoon red pepper flakes

2 tablespoons low-sodium tomato paste

1 tablespoon fresh oregano, chopped,
  or 1 teaspoon dried oregano

1 tablespoon balsamic vinegar

1 tablespoon fresh basil, chopped

¼ teaspoon salt

freshly ground pepper to taste

directions

1. Pour jar of crushed tomatoes into blender. Add the sliced tomatoes and pulse lightly. Set aside.
2. Bring vegetable broth to a simmer in a large skillet over medium-high heat.
3. Add onion, garlic, oil, and red pepper flakes. Cook until onion is translucent (about 5–7 minutes).
4. Add tomato paste and cook 1 minute while stirring.
5. Reduce heat to medium, stir in tomato mixture and dried oregano (if using fresh herbs, do not add yet), and cook about 15 minutes to blend flavors, stirring occasionally to make sure mixture doesn't stick to pan.
6. Remove from heat and let cool for a few minutes. Add tomato mixture to blender, along with vinegar, basil, oregano (if using fresh herbs), salt, and pepper. Pulse lightly to combine. Serve warm or refrigerate until ready to serve. Leftovers can also be stored in the freezer.

* It's important to purchase **olive oil that comes in a dark glass container**, as light exposure degrades the oil. Also, store your oil in a dark, cool cabinet rather than on a countertop, which is likely brighter and warmer.

texas bbq sauce

prep time: 5 minutes • **cook time:** 5 minutes • **makes:** 2½ cups

We call for this sauce in the Sweet Potato Chickpea BBQ Wraps (page 99) and also in the Pulled "Pork" Sliders (page 94), but go ahead and get creative by adding this to any meal!

Note: If you are making this for the Pulled "Pork" Sandwiches, you will definitely have leftover sauce. Refrigerate the rest to have with other meals throughout the week. Because it contains a good amount of apple cider vinegar, this sauce will last up to 10 days in the fridge.

ingredients

2 cups tomato sauce

¼ cup maple syrup

2 tablespoons regular molasses (not blackstrap)

2 tablespoons apple cider vinegar

3 tablespoons Worcestershire sauce

2 tablespoons liquid smoke

1½ tablespoons chili powder

1 tablespoon garlic powder

½ teaspoon black pepper

¼ teaspoon salt

½ teaspoon ground coffee or espresso

directions

1. Add all ingredients to a blender or food processor and blend until smooth.
2. Add to a pot over medium heat. Once the sauce begins to simmer, cook for 3–5 minutes, stirring until spices cook through and it is warmed.
3. Use immediately or store in refrigerator.

alfredo sance

pre-prep: *Soak cashews and almonds for 4 hours or overnight, or quick soak (page 12).*

If using dried chickpeas, soak for 6 hours or overnight and cook for 1 hour (page 11)

prep time: *15 minutes* • *cook time*: *2–3 minutes* • *makes*: *3 cups or 5–6 servings*

Because it's typically made with heavy whipping cream, butter, and parmesan cheese, you're used to skipping Alfredo pasta on the menu if you're a plant-based eater or just trying to be healthy. But fear not, we have created a rich Alfredo Sauce recipe with the same delicious taste and texture that also provides lots of protein and nutrients from the chickpeas, nutritional yeast, nuts, and seeds. This tasty sauce can be used with our Sweet Potato Gnocchi (page 111), other pastas, and even spaghetti squash.

ingredients

1 tablespoon olive oil

4 cloves garlic, minced

½ cup cashews, soaked and drained

½ cup almonds, soaked and drained

4 tablespoons pine nuts

1 teaspoon miso sauce

½ cup nutritional yeast

½ cup chickpeas, cooked or canned, drained

1 teaspoon salt

1 teaspoon Dijon mustard

½ teaspoon onion powder

1 cup water

½ cup plant-based milk + 1–2 tablespoons
set aside to thin sauce if needed

1½ teaspoons lemon juice

1 teaspoon lemon zest

pinch of nutmeg & black pepper

serving ideas

Sweet Potato Gnocchi (*page 111*) • other pasta of choice • spaghetti squash

* **Pine nuts** *were and continue to be an important staple for Native Americans, particularly the Washo, Shoshone, Paiutes, and Hopi people.*

directions

1. Heat ½ tablespoon olive oil in a small pan. Add minced garlic and sauté for 3-4 minutes. Add 2 tablespoons of the pine nuts and sauté another 2 minutes.

2. In a blender, puree cashews, almonds, sautéed garlic/pine nuts, miso, nutritional yeast, chickpeas, garlic, sea salt, Dijon, onion powder, water, milk, and lemon juice until very smooth. If sauce is too thick, add 1–2 tablespoons of plant-based milk.

3. Heat the other ½ tablespoon olive oil in small pan. Sauté remaining pine nuts for 2–3 minutes until they start to brown.

4. Pour sauce in medium pot and heat on low for 2–3 minutes. Stir in lemon zest, nutmeg, pepper, and remaining sautéed pine nuts.

5. Serve sauce with pasta of your choice. Feel free to sprinkle on nutritional yeast for extra flavor and nutrition!

cheesy cheddar sauce

prep time: 5 minutes • **cook time:** 5–6 minutes • **makes:** 1 cup

Here we have our super easy non-dairy, cheesy sauce! Use this sauce for Sweet Potato Enchiladas (page 112), Sweet Potato Burger or Black Bean Burger (page 127), nachos, or as a dip for veggies.

ingredients

¾ cup coconut milk, canned

3 tablespoons nutritional yeast

2 tablespoons tapioca flour

½ teaspoon salt

½ teaspoon onion powder

¼ teaspoon garlic powder

pinch of smoked paprika

directions

1. Put all ingredients into a saucepan and stir well.

2. Once everything is mixed, turn the heat on and bring mixture to a boil while stirring.

3. Let simmer on low for about 1 minute.

4. Serve immediately or store in fridge.

almond tofu ricotta cheese

pre-prep: Soak almonds for either 6 hours or overnight (for quick soak see page 12)
prep time: 25 minutes • *serves*: 8

This combination of ingredients creates the same light and creamy texture as ricotta cheese. Spread this on our Gluten-Free Bagels (page 40), add to pasta, or use as a dip for crackers or veggies!

ingredients

1 teaspoon olive oil

1 clove garlic, minced

1 cup almonds, soaked and drained

3½ ounces extra-firm tofu, drained

½ cup unsweetened soy milk

4 tablespoons plant-based yogurt, plain
*We recommend coconut-based yogurt
for the most neutral taste.*

2 teaspoons nutritional yeast

1 teaspoon lemon juice

½ teaspoon dried parsley

½ teaspoon salt

a few basil leaves

directions

1. Add olive oil to small pan and heat on medium. Add garlic and sauté for 3-4 minutes.

2. Add all ingredients to a blender or food processor and mix until smooth.

3. Serve immediately or store in refrigerator. Can also be stored in the freezer.

cashew parmesan cheese

pre-prep: Soak cashews for either 4 hours or overnight, or quick soak (page 12)
bake time: 1 hour • **prep time**: 5 minutes • **makes**: ¾ cup

This cheese is extremely versatile and is great to have as a staple in your kitchen. We specifically call for it in our Pepperoni Pizza Pinwheels (page 84) and Pesto Pasta (page 89), but feel free to sprinkle it on anything that could use a little cheesiness!

ingredients

¾ cup cashews, soaked and drained

3 tablespoons nutritional yeast

½ teaspoon salt

¼ teaspoon garlic powder

directions

1. Preheat oven to 200°F. Place soaked cashews on a baking sheet and bake for 1 hour until thoroughly dry.
2. Place cashews in a blender or food processor along with the rest of the ingredients and mix/pulse until a fine parmesan-like texture is achieved.
3. Store in the refrigerator to keep fresh. Lasts for several weeks.

* Nutritional yeast

(also known as nooch in the vegan community) is high in vitamin B12 and is also a complete protein, containing all nine amino acids. Vegans love nooch because it does a great job of mimicking the flavor of cheese. You can put it on anything, including pasta, popcorn, sauces, or even salads.

mozzarella cheese

pre-prep: Soak cashews for 4 hours or overnight, or quick soak (page 12)
prep time: 10 minutes • **cook time:** 10 minutes • **makes:** 1 cup

Melty, gooey, stretchy mozzarella! If you're trying to be plant-based but miss sinking your teeth into creamy, thick cheese, this recipe will be your go-to. We use this cheese in our Sloppy Joes (page 115) and our Pepperoni Pizza Pinwheels (page 84).

ingredients

½ cup cashews, soaked and drained
Cashews contain protein and are one of the few food sources that are high in copper.

1¼ cups water

¼ cup tapioca flour

2 tablespoons nutritional yeast

2 teaspoons lemon juice

½ teaspoon onion powder

½ teaspoon garlic powder

¾ teaspoon salt

directions

1. Add cashews and all other ingredients to blender and blend on high speed until smooth and creamy (1–2 minutes). Mixture will be very thin, but will thicken when cooked.

2. Add mixture to a small pot. On medium heat, stir mixture for about 5–10 minutes, until thick. Serve immediately or store in refrigerator.

cashew mushroom gravy

pre-prep time: Pre-soak cashews overnight • **prep time:** 15 minutes
cook time: 10 minutes • **makes:** 4 cups or 8–10 servings

Once you see how easy it is to make this delicious gravy, you will never want to buy store-bought gravy again! The rich combination of cashews, mushrooms, and spices will elevate anything you use it to douse. We feature it in our Loco Moco (page 80), but it is also amazing over Garlic Cauli/Potato Mash (page 153) and Miso Meatballs (page 83).

ingredients

1 teaspoon olive oil

3 cloves garlic, minced

1 shallot or ¼ onion, diced

10 ounces baby bella mushrooms,
   chopped (*small portobello mushrooms*)

pinch salt

½ cup red wine

1 cup cashews, soaked and drained

2 cups vegetable broth + ½ cup
   more for thinning the sauce

3 tablespoons tamari or coconut aminos

2 tablespoons cornstarch

2 tablespoons nutritional yeast

2 teaspoons onion powder

1 teaspoon garlic powder

½ teaspoon salt

½ teaspoon thyme

½ teaspoon rosemary

directions

1. Add olive oil to large pan over medium-high heat. Add garlic and shallot or onion and cook until slightly browned.
2. Add mushrooms and a pinch of salt, and cook until mushrooms are browned.
3. Add red wine and stir. When mushrooms are slightly browned, set aside.
4. Combine cashews, 2 cups of veggie stock, tamari, cornstarch, nutritional yeast, onion powder, garlic powder, and salt in a blender. Process until smooth.
5. Transfer mixture to medium pot, and add mushroom mixture, thyme, and rosemary. Cook uncovered on medium/low heat for 8–10 minutes. Slowly add in up to ½ cup vegetable broth, until sauce reaches your desired consistency.

We use baby bella mushrooms,
but feel free to try crimini, white button, or other types of mushrooms.

cauliflower / cashew sour cream

pre-prep: Soak cashews for 4 hours or overnight, or quick soak (page 12)

prep time: 20 minutes plus 1 hour in refrigerator before serving • *makes*: 1½ cups

We love the fact that eating plant-based can still mean getting to have all of your favorite meals and condiments. It's just about getting creative (yes, cauliflower can actually make for a delicious, creamy sour cream)! This recipe is a great addition to so many meals, including Dad's Hearty Chili (page 68), Sweet Potato Enchiladas (page 112), and Sweet Potato Hash Browns (page 169).

ingredients

¾ cup cauliflower florets

1 cup cashews, soaked and drained

½ cup coconut yogurt, plain and unsweetened

1½ tablespoons lemon juice

1½ teaspoons miso paste

1 teaspoon apple cider vinegar

⅛ teaspoon salt

directions

1. Place cauliflower florets in a pot with a steamer basket and steam for 6–8 minutes (see page 12 for steamer directions). Cool for 10 minutes.
2. Place cauliflower in a blender along with all other ingredients.
3. Blend ingredients on high for 1 minute.
4. Refrigerate for 1 hour and serve.

ranch dressing

pre-prep: Soak cashews for either 6 hours or overnight, or quick soak (page 12) • *prep*: 10 minutes plus 30 minutes in refrigerator before serving

makes: 1 cup

ingredients

½ cup plain yogurt, plant-based

½ cup cashews, soaked and drained

¼ cup coconut milk or other plant-based milk

1½ tablespoons apple cider vinegar or white vinegar

1 tablespoon red wine vinegar

1 teaspoon onion powder

1 teaspoon garlic powder

½ teaspoon dried dill

¼ teaspoon dry mustard seed

¾ teaspoon salt

directions

1. Mix all ingredients on high in a food processor for 1 minute. Store in the refrigerator for a few hours to thicken.

2. If dressing is too thick, add water and apple cider vinegar until the desired consistency is achieved. Stir or shake before serving.

* Cashews are great for any recipe that requires a creamy substitute for dairy. They contain healthy fats and antioxidants such as copper, magnesium, and manganese.

dressings

If you ever look at the back of store-bought dressings, you will notice all kinds of oils, artificial flavors, preservatives, and other ingredients you can't pronounce. Because they're staples in many American households, we decided to tweak these common salad dressings and make them not only plant-based, but full of whole, healthy ingredients. We feature the Ranch Dressing with our Southwest Salad (page 96) and the Caesar Dressing with our Sweet Potato Chickpea BBQ Wraps (page 99). Our Thousand Island Dressing is great on salads too, or as a topping for the burgers (page 127).

thousand island dressing

pre-prep: Soak cashews for 4 hours or overnight, or quick soak (page 12) • *prep time*: 10 minutes
before serving: Place in refrigerator for 30 minutes • *makes*: 1 cup

ingredients

½ cup cashews, soaked and drained
½ cup water (*separate from soaking cashews*)
¼ cup + 2 tablespoons ketchup
1 tablespoon vegan mayonnaise
1 tablespoon fresh lemon juice
2 tablespoons diced onion
¼ teaspoon paprika
¼ teaspoon dried mustard
¼ teaspoon Himalayan salt
2 tablespoons sweet pickle relish
(1 tablespoon for blending and 1 for stirring in after)

directions

1. Add cashews and all other ingredients to blender. Save 1 tablespoon relish for adding after blending. Blend on high for 1 full minute.
2. Pour into container and stir in remaining tablespoon relish. Serve immediately and store remaining dressing in refrigerator.

caesar dressing

prep time: 10 minutes
before serving: Place in refrigerator for 30 minutes
makes: 1 cup

ingredients

1 tablespoon olive oil
2 large garlic cloves, minced
½ heaping cup pine nuts
¼ cup + 3 tablespoons water
2 tablespoons lemon juice
3 tablespoons nutritional yeast
½ tablespoon red wine vinegar
½ tablespoon apple cider vinegar
½ teaspoon dried parsley
½ teaspoon salt
¼ teaspoon pepper

directions

1. Heat olive oil in a small frying pan. Add garlic and sauté for 4–5 minutes.
2. Add garlic and all other ingredients to a small food processor and blend until very smooth and creamy. Place in refrigerator for 30 minutes.

* *We call for this dressing in the BBQ Chickpea Wraps (page 99) but also recommend it* **with any salad!**

desserts

mel's favorite black bean avocado brownies

pre-prep: If using dried black beans, soak for 6 hours or overnight and cook for 1 hour (page 11). 10 minute soak for dates.

prep time: 25 minutes • **bake time**: 25–30 minutes • **makes**: 9–12 brownies

My sister Mel's favorite dessert in the world is a warm, fudgy chocolate brownie with vanilla ice cream. She insisted that we come up with a great brownie recipe for the book, taking highly nutritious ingredients and transforming them into the fudgy and delectable brownie we all know and love. We know what you're thinking ... *black beans and avocado, really?* Trust us on this one. As soon as it comes out of the oven, Mel is the first one in line with her bowl, adding a big dollop of vanilla cashew ice cream. Before baking, she also loves adding shredded coconut to the top! *mackenzie*

ingredients

4 dates, soaked in water for 10 minutes, drained

1 cup black beans, cooked or canned, drained

2 flax eggs (*page 10*)

½ of a large avocado

1 tablespoon coconut oil, melted,
    + a little extra for greasing pan

½ cup + 1 tablespoon unsweetened cacao powder

½ teaspoon baking powder

¼ teaspoon baking soda

¼ teaspoon salt

1 teaspoon vanilla extract

½ cup coconut sugar

1–4 teaspoons water (*depending on consistency*)

⅓ cup chocolate chips + 2 tablespoons (*for topping*)

½ cup walnuts, chopped

topping ideas

1 tablespoon unsweetened shredded coconut

directions

1. Preheat oven to 350°F. Grease an 8 x 8" baking pan with coconut oil. Soak dates in water for 10 minutes.

2. Place all ingredients except for chocolate chips, walnuts, and water in a blender or food processor. Puree until mixture forms a smooth batter. If batter is too thick, add 1–4 teaspoons water.

3. Stir in chocolate chips (except those reserved for the topping) and walnuts.

4. Pour batter into prepared pan, sprinkle with remaining chocolate chips and shredded coconut (optional).

5. Bake for 25–30 minutes. (Just a heads up, the texture will be gooey. If you prefer a firmer texture, cook an extra 10 minutes.)

6. Cool and enjoy!

* *These high-fiber beans have 15 grams of fiber per cup and also provide an extra boost of benefits by supplying protein, magnesium, manganese, and folate.*

chocolate chip cookies

prep time: 10 minutes • *bake time*: 13 minutes • *makes*: 12–14 cookies

Super simple yet so delicious—if you've been searching for the best vegan, gluten-free cookie recipe, you can stop looking now! The combination of almond and oat flour not only serves up more nutrition than you'll find in the average cookie, but also provides incredible flavor.

ingredients

1 cup oat flour

1 cup almond flour

½ cup coconut sugar

½ teaspoon baking soda

½ teaspoon baking powder

½ teaspoon salt

⅓ cup coconut oil, melted

3 tablespoons plant-based milk

   + 1 tablespoon extra

   if needed (*We use almond*)

2 flax eggs (*page 10*)

2 teaspoons vanilla extract

⅓ cup dark chocolate chips

directions

1. Preheat oven to 350°F. Line a baking sheet with parchment paper.
2. In a large bowl, combine oat flour, almond flour, coconut sugar, baking soda, baking powder, and salt. Mix until thoroughly combined.
3. In a medium bowl combine melted coconut oil with almond milk, flax eggs, and vanilla extract. After mixing well, pour wet ingredients into the large bowl of flour mixture and mix together. If dough seems crumbly, add an extra tablespoon of almond milk. Add chocolate chips to the mixture and stir.
4. Scoop out about 1½ tablespoons dough for each cookie, and arrange on the baking sheet. Slightly flatten dough with a spoon. Bake for 12–13 minutes, until edges are light golden brown. Let cool and serve.

chocolate pie

pre-prep: Soak cashews for 4 hours or overnight, or quick soak (page 12)
crust: *prep time*: 10 minutes • *bake time*: 20 minutes
filling: *prep time*: 20 minutes (while crust is baking and cooling)
bake time: 45 minutes • *before serving*: Place in refrigerator for 1 hour

Before we set out to write this book, the thought of baking a chocolate pie, let alone a delicious gluten-free, vegan pie, felt like a daunting task. We somehow figured it out, and we are excited to share it with you because, like so many recipes that seem intimidating, it was really a piece of cake (pie)! And, you may find it hard to believe that this incredibly rich chocolate mousse filling is packed with protein from the sneaky addition of tofu. (Yes, you read that right!)

ingredients

crust

1 cup cashews, soaked and drained

6 tablespoons coconut oil, melted
   + a bit for coating pie dish

6 tablespoons maple syrup

2 teaspoons vanilla extract

½ teaspoon salt

2 cups gluten-free flour + extra for hands

filling

¾ cup dark chocolate chips

12 ounces firm tofu, drained

½ cup maple syrup

¼ cup plant-based vanilla yogurt

½ cup cacao powder

1½ teaspoons vanilla extract

¼ teaspoon salt

topping ideas

coconut whipped cream
   We recommend the So Delicious CocoWhip.

* *It's hard to believe that **a block of tofu** is one of the key ingredients in this pie. Tofu is packed with protein, contains all of the essential amino acids your body needs, and is also a good source of calcium, manganese, copper, selenium, and phosphorus.*

directions

1. Preheat oven to 350°F. Lightly coat a 9" pie dish with coconut oil.

2. In a blender or food processor, blend cashews until finely ground.

3. For the crust, mix oil, maple syrup, vanilla, and salt in a medium bowl. Add ground cashews and flour, and mix well.

4. Dish mixture into pie dish. Lightly dust hands with the gluten-free flour and press mixture evenly to the bottom and up the sides. Form a nice, even rim around the top of the pie pan.

5. Bake for about 20 minutes, until crust is golden brown. Remove from oven and let cool for 20 minutes while making filling.

6. To prepare filling, decrease the temperature to 325°F. Melt chocolate chips in a small saucepan on low on stovetop.

7. In a blender, blend tofu, maple syrup, yogurt, cacao powder, vanilla, and salt until smooth. Add in melted chocolate and blend for 10 seconds.

8. Pour pie mixture into pie crust. Bake for 45 minutes.

9. Refrigerate the pie until it is cold. Top with coconut whipped cream.

leah PENNIMAN

Leah Penniman, 2019 recipient of the James Beard Foundation Leadership Award, has dedicated her life to tending the soil and advocating for an anti-racist food system.

Penniman has worked at the Food Project, Farm School, and Many Hands Organic Farm; she co-founded Youth Grow, and has worked with farmers internationally in Ghana, Haiti, and Mexico. In 2010, Penniman co-founded Soul Fire Farm in Grafton, New York, a Black and Indigenous-led community farm dedicated to ending racism in the food system and training the next generation of activist farmers. Her book *Farming While Black* is a love song for the earth and her people.

LIVING IN A FOOD APARTHEID

When Leah Penniman had a hard time fitting in at school, the land became her salvation, and the forest became her best friend. She would garden and grow strawberries with her beloved grandmother, Brownie McCullough.

As a teenager, working at a farm called the Food Project, she was inspired by the intersection between the natural environment and social justice.

When her children were young, Leah struggled to find healthy food for them. She explained that her only options were a corner store, a liquor store, and a McDonald's. **"Our community is considered a food desert by the USDA but we use the term 'food apartheid' because this is a human-created system of segregation, not a natural ecosystem."**

With a mission to end racism in the food system, Soul Fire Farm is an 80-acre community farm. This BIPOC (Black, Indigenous, and People of Color) centered community works to increase the leadership of people of color in the food justice movement.

"There are a whole host of entry points when it comes to food," Penniman states, and her organization opens as many entry points as possible.

Centered in the Albany area, the farm serves the surrounding community in many ways: by providing low-cost food, sharing skills on sustainable agriculture, and inspiring spiritual activism, health, and environmental justice.

Through "Afro-Indigenous farming" immersion and workshops, Soul Fire Farm equips hundreds of members with the land-based skills needed to reclaim leadership as farmers and food justice organizers. The graduates receive ongoing mentorship to access resources, land, and training to amplify their voice in the food system.

Penniman's mission also focuses on healing the communities' relationship to the earth itself. "It's a pretty natural remembering for people," she says as she speaks about honoring ancestors. To regard those who built the agricultural system of this country on their backs, Penniman holds ceremonies and asks permission before planting. She brings in African cosmology in order to carry on ancestral ways of respecting the Earth.

Penniman goes beyond local change and also works in the policy space. In March 2020, I collaborated with Penniman on a memo titled Land Access For Beginning and Disadvantaged Farmers. Together, we highlighted the dangerously low percentage of farms owned by people of color, called out the discriminatory practices by the USDA, and proposed policy ideas for the Green New Deal. This effort was intended to eradicate the historical discrimination and increase support for socially disadvantaged farmers. **In 1910, black farmers made up 14% of America's farmers. Today, that figure is less than 2%.** See the Land Access food fact on page 63 for more information.

In 1910, Black farmers made up 14% of America's farmers. Today, that figure is less than 2%.

> "We couldn't rely on a system that is a capitalist system and a racist system. To free ourselves, we must feed ourselves." With a mission to end racism in the food system, Leah is feeding us all.

LEAH PENNIMAN'S

haitian sweet potato bread

prep time: 20 minutes if using a hand grater or less with a food processor • *cook time:* 1½ hours

ingredients

2 pounds sweet potatoes, peeled

1 large banana, cut into 1-inch pieces

1 cup brown sugar, 1 tablespoon reserved

½ cup raisins

1 teaspoon grated ginger

¼ teaspoon salt

12 ounces evaporated milk

(or ½ cup honey and ½ cup coconut oil)

1 teaspoon vanilla extract

3 eggs

½ teaspoon grated nutmeg

1 teaspoon ground cinnamon

zest of 1 lemon

1½ cups coconut milk

3 teaspoons melted butter *(or coconut oil)*

To make it plant-based:
- flax eggs can be substituted for eggs (page 10)

directions

1. Preheat the oven to 350°F.
2. Grate the sweet potatoes into a large bowl and mash the banana into the sweet potatoes.
3. Add the rest of the ingredients (except the 1 tablespoon brown sugar) and mix until the batter is fully incorporated with a pudding-like consistency.
4. Spread evenly into a 9×13" baking pan and evenly sprinkle the remaining brown sugar over the top.
5. Bake for 1½ hours or until a toothpick inserted into the center comes out clean. Cool completely before serving.

Leah tending to the crops at Soul Fire Farm

distant cousins *(From page 116 of Farming While Black)*

Many of the crops we have curated and cherished over millenia do not thrive in the soil to which we have been transplanted. For example, the yam (Dioscorea spp.), "King of Crops," sits tired looking on the shelf of the African market in town, having endured a transatlantic journey. Our ancestors recognized that the sweet potato, of the Convolvulaceae family, native to the Americas, could supplant the nutritional and spiritual role of the yam in Black agrarian life.

planting and harvest rituals *(From page 62 of Farming While Black)*

A pile of newly harvested sweet potatoes, the yams of the Diaspora, rested on a white sheet to the east of our gathering space. We had harvested these yams from the cool soil to the rhythm of a drum, as is the custom at the start of Manje Yam, the Haitian festival for the eating of new yam ... Banana leaves are selected because they perpetually self renew from the roots, representing the eternal nature of the Divine ... Then we lay down on the banana leaves and rolled ourselves toward the yams in the East, the land of Ginen, our ancestral home where the Loa would fortify us spiritually for the year ahead. After the ritual return back across the Middle Passage to receive the blessings of our ancestors, we could prepare and eat the new yam ... The veneration of the new yam marks the beginning of the harvest season. It is a time to give thanks to the Spirit of the Land for a good yield and successful harvest.

Enjoy this delicious gluten-free dessert from our homeland.

Dear Grandmommy,

Thank you for being the first person to teach me how to coax nourishment from the earth. I remember my small hands next to your worn and tired hands harvesting strawberries for jam and pies. As a daughter of the great migration, you brought stories and recipes from the Carolinas and passed onto me a reverence for sweet potato, greens, and long beans. Now, I play with your recipes and teach them to my own children, and on to theirs. May our circles be unbroken.

Love,

leah

tahini banana bread

prep time: 15 minutes • **bake time:** 35–40 minutes • **makes:** One loaf or 9 squares

We never knew how delicious the combination of bananas and tahini were until we created this rich, nutty bread, great for either breakfast or a mid-morning snack. It also can be made into squares for an incredible dessert, simply by swapping out the loaf pan for a baking dish, and topping it with some vanilla coconut ice cream! People are often shocked when we tell them there is no real butter in this dish, and it is not only vegan, but also gluten-free. Tiger nut flour, made from a small, highly nutritious root vegetable, provides an amazing nutty taste. This incredible bread is destined to become a family favorite!

ingredients

2 ripe bananas

2 flax eggs *(page 10)*

¾ cup drippy tahini butter
We recommend Mighty Sesame Co. for squeezable tahini.

¼ cup maple syrup

1 teaspoon vanilla extract

¼ cup coconut oil, melted

1 cup tiger nut flour *(or almond flour)*

2 tablespoons coconut flour

¼ teaspoon salt

1 teaspoon baking soda

1 teaspoon cinnamon

½ cup chocolate chips or gems

1 teaspoon flaky salt for
topping before baking

directions

1. Preheat oven to 350°F. Grease or line a loaf pan or an 8x8" baking dish.
2. Mash bananas in a bowl with fork.
3. Mix in flax eggs, tahini, maple syrup, vanilla extract, melted coconut oil.
4. Add in tiger nut flour, coconut flour, salt, baking soda, cinnamon, and mix.
5. Fold in chocolate chips.
6. Pour batter into baking dish and spread out evenly. Top with flaky salt.
7. Bake for 30–35 minutes (35–40 if making loaf). Turn oven off, open door slightly, and let the pan sit in the oven for 5–10 minutes. Let cool before serving.

* **Tiger nuts** *were one of the first plants cultivated in Egypt and traditionally used as both food and medicine.*

chickpea blondies

pre-prep: If using dried chickpeas, soak 6 hours or overnight and cook 1 hour (page 11)
prep time: 10 minutes • bake time: 20 minutes • makes: 12 Blondies

We know what you're thinking ... chickpeas, *really*? Yes, really! Chickpeas are an excellent ingredient for this recipe, maintaining the cake-like consistency of blondies while adding fiber and protein. The combination of chickpeas, nut butter, and bananas creates a chewy, creamy, delicious (yet secretly healthy) dessert.

ingredients

coconut oil for pan

1 cup chickpeas, cooked or canned, drained

½ cup oats

⅓ cup plant-based vanilla protein powder

¼ cup nut butter

2 ripe bananas

¼ cup coconut milk

¼ cup maple syrup

1 teaspoon vanilla extract

½ teaspoon baking powder

½ teaspoon baking soda

¼ cup chocolate chips

directions

1. Preheat the oven to 350°F.
2. Grease an 8x8" pan with coconut oil.
3. In a blender, add all of the ingredients except chocolate chips, and blend until smooth. After blending, stir in half of the chocolate chips.
4. Pour the mixture into baking pan, and sprinkle the remaining chocolate chips on top.
5. Bake for 20 minutes.
6. Let pan cool for 10 minutes. Serve blondies warm. Leftovers can be refrigerated or stored in the freezer.

*Chickpeas have been a part of certain traditional diets for over 7,500 years.

chocolate date caramel tart

prep time: 1 hour • *bake time*: 30 minutes • *before serving*: Place in refrigerator for 1 hour • *serves*: 8

Calling all chocolate lovers ... this one's for you! While it takes a bit of time to prep and cool, the recipe is actually pretty simple, and did we mention it tastes incredible? While the typical chocolate tart calls for heavy cream, milk, eggs, and wheat flour, this recipe uses all plant-based, gluten-free ingredients to produce a decadent filling surrounded by a buttery, shortbread-like crust. We suggest pairing this recipe with coconut whip or plant-based vanilla ice cream. A perfect dessert for special occasions.

ingredients

crust

2 cups almond flour

⅛ teaspoon salt

¼ cup maple syrup

¼ cup coconut oil, melted

walnut filling

10 Medjool dates soaked in ½ cup warm water for 10 minutes

2 tablespoons coconut oil, melted

dash of salt

1 teaspoon vanilla extract

¾ cup + ¼ cup toasted walnuts or pecans, chopped

ganache

½ cup maple syrup

¼ cup coconut oil

½ cup unsweetened cacao or cocoa powder

Recipe continued on page 208

directions

crust

1. Grease a 9" round pan.
2. Place almond flour and salt into a large bowl.
3. Add maple syrup and melted coconut oil and blend together with a fork for 2 minutes, until it resembles a coarse meal.
4. Press firmly and evenly into the bottom and sides of the pan with fingers.
5. Place in the refrigerator for 20 minutes before baking.
6. Preheat oven to 350°F. Bake for 15 minutes, until golden brown.
7. Remove from oven and set aside to cool while making filling.

filling

1. In a blender or food processor, puree the dates along with the soaking liquid, melted coconut oil, salt, and vanilla extract until thick and smooth.
2. Stir in ¾ cup of chopped toasted nuts. Spread mixture evenly into the pan over the crust and bake for 15 minutes at 350°F. Cool.

ganache

1. In a blender or food processor, blend together maple syrup and melted coconut oil until well combined.
2. Add cacao or cocoa powder and blend until smooth. Pour over the date-nut base and spread out evenly.
3. Return pan to fridge for 1 hour before serving.
4. Sprinkle with extra salt and remaining chopped toasted nuts before serving.

We love using dates as a substitute for white sugar because of the nutrients they provide such as potassium, magnesium, copper, and vitamin B6, as well as fiber and antioxidants.

chocolate hazelnut spread

pre-prep time: 15 minutes • **bake time:** 12–15 minutes • **prep time:** 10 minutes

The typical hazelnut spread you put on toast or in crepes, like Nutella, contains palm oil in addition to artificial flavors and sugar. The harvesting of palm oil is a major driver of deforestation in some of the world's most biodiverse forests, destroying the habitat of endangered species such as the orangutan, pygmy elephant, and Sumatran rhino. Why support that when you can make your own hazelnut spread, with the same silky texture and unbelievable nutty flavor? We recommend using this spread for dipping strawberries and bananas, drizzling over ice cream, and spreading on toast.

ingredients

2 cups hazelnuts

1 tablespoon baking soda

1 cup chocolate chips

1 tablespoon coconut oil

½ teaspoon vanilla bean powder

pinch of salt

directions

1. Preheat oven to 350°F. Boil hazelnuts in a small pot of water with 1 tablespoon baking soda for a few minutes. Remove hazelnuts and immerse them in cold water, then peel skins off. Line baking sheet with parchment paper.

   The best way to peel off the skins after boiling them is to place a handful of nuts into a slightly dampened kitchen towel and rub them together. Place peeled nuts in a separate bowl and continue the process until all nuts have gone through this process. Some skins will remain on, which is fine.

2. Bake hazelnuts for 12–15 minutes. Remove from oven and cool.

3. Add hazelnuts to blender or small food processor and blend until they turn into hazelnut butter. Hazelnuts will first turn into flour, then butter. This takes 5–7 minutes.

4. Add chocolate chips and coconut oil to a small saucepan and melt on low heat, stirring continuously to blend.

5. Add chocolate mixture to blender or food processor, along with vanilla powder and salt. Continue to blend until mixture is smooth and creamy.

6. Serve and enjoy! Store leftover spread in the refrigerator.

✱ Hazelnuts, also known as filberts, are rich in unsaturated fats, manganese, copper, antioxidants, and vitamin E.

DR. *steven* LAWENDA

Steven Lawenda, MD., is a board-certified Family Medicine physician at Kaiser Permanente in Southern California. In 2013, Dr. Lawenda transformed his own health with plant-based nutrition.

Since then, he has co-created Lifestyle Medicine programs emphasizing food as medicine that have transformed both the lives of patients and the way in which healthcare is delivered at his medical center. Lawenda has co-authored a review of Lifestyle Medicine with several esteemed colleagues that was published in the *Permanente Journal*. He is a two-time recipient of the Southern California Permanente Medical Group Everyday Hero award as well as a winner of the Kaiser Permanente Antelope Valley Physicians' Exceptional Contribution Award.

Dr. Steven Lawenda learned about the power of whole-food, plant-based (WFPB) nutrition to prevent and even reverse and cure disease almost by accident. On a long drive home from a family vacation, his wife, Patty, turned on the audio book *Eat To Live* by Joel Fuhrman (a board-certified medical doctor), which completely transformed Lawenda's way of thinking. The revelations around the power of whole-food, plant-based nutrition to prevent and reverse diseases from which millions are suffering was backed up by years of medical studies, and Lawenda was shocked that he had never learned any of this in his years of medical school. Right then and there, Lawenda made the decision to switch to a WFPB diet.

This decision changed his life. He lost 75 pounds within eight months and his prediabetes, acid reflux, and fatty liver seemed to vanish. Because he felt so much better, he decided to incorporate this diet into his clinical practice. He now recommends a WFPB diet to almost every patient struggling with a chronic illness, and he has seen clear improvements in almost all of them. Lawenda believes that more than 70% of chronic illnesses can be either reversed or prevented with whole-food, plant-based nutrition.

He is now one of the major voices supporting The Plantrician Project, an endeavor to transform human health, health care, and the food system by promoting a WFPB diet to patients and getting physicians worldwide to do the same.

One of the hardest parts of Lawenda's lifestyle change was the cravings. **"There are a lot of studies now that show that processed foods can be just as addicting as drugs,"** he says. "When you stop eating processed foods and animal foods, cravings are normal. A lot of people don't realize that that part is just temporary," which is often the reason why people don't adopt this lifestyle.

When Lawenda went plant-based, underlying issues within the food industry came to light. He realized that highly processed, addictive, and toxic foods were hugely profitable, and that true change requires a grassroots shift in demand for healthier products. Lawenda also spoke about how food is a powerful medicine and unfortunately, many health care providers are unaware of the benefits that come with a WFPB diet. He was frustrated that people are not taught this valuable information and that the healthcare system does not facilitate this transfer of knowledge. Thus, he is passionate about the free programs he has created, as they provide much needed resources about the power of plant-based eating.

Lawenda's introductory class, open to everyone, often draws a large audience. In class, he reviews scientific studies demonstrating the benefits of a plant-based diet, shows videos of patient success stories, gives practical tips to start the journey, and shares fun recipes that attendees can try. Lawenda also offers a 14-week program open to those with chronic conditions who need more intense education and support.

Lawenda believes that more than 70% of chronic illnesses can be either reversed or prevented with whole-food, plant-based nutrition.

Included in his list of medical success stories are patients who no longer need amputations due to diabetes complications, patients who no longer need insulin, and those who no longer need weight-loss surgery. One patient struggled with chest pain and tested a diet with less red meat and more chicken and fish. When that didn't help, he was told that he needed bypass surgery. "I happened to be his doctor," says Lawenda, "and mentioned we had this class on the value of plant-based eating." After attending Lawenda's class, the patient decided to try going plant-based, hoping it could reverse his heart disease. Sure enough, within two months, his chest pains were gone.

Dear Patty,

Thank you so much for your insistence that we listen to the audiobook of *Eat to Live* during our car ride home from vacation when I preferred to relax with music. And an even bigger thank you for supporting and, in fact, joining me on my transformative journey to a whole-food plant-based lifestyle. This journey, with your encouragement and support, has transformed, if not saved, my life. Likewise it has transformed and saved my medical career as well as the lives of numerous patients I am privileged to serve. I am forever grateful.

Love, your husband,

steve

Lawenda notes that people generally don't acknowledge the powerful influence, good and bad, that food can have. "When you look at that list of all the other killers," he says, "there's a stigma attached to a lot of those [such as] drugs, alcohol, cigarettes, but there's no stigma [on food]. In fact, it's the opposite of a stigma. We celebrate comfort food like hot dogs, ice cream, pizza, and fried chicken ... There was an article in *USA Today* where they were ranking the top hamburgers around the country. And I'm thinking, this is in *USA Today*, one of the top, most popular newspapers. They know that this is going to be of interest to people. You don't see an article where they say, here's where you get the best heroin ... there's just such a disconnect between food and the power behind it."

Though it took some time, says Lawenda, "our leadership at Kaiser is supportive of what I'm doing. The medical director of my medical center actually became plant-based himself after a few people saw my transformation ... And once we started and we showed them some data that this kind of diet is really working, then they became more supportive." Now, hundreds of people attend Lawenda's workshops and he has gained strong recognition for the attention he has brought to the health benefits of plant-based eating. As a well-respected doctor, he has the platform to be an influential changemaker who can create a meaningful shift toward healthier diets.

DR. STEVEN LAWENDA'S
banana cacao "nice cream"

pre-prep time: 5 minutes including blending • **Freeze time:** 6 hours

makes: 2 cups, 4 servings

ingredients

4 ripe bananas

1 teaspoon vanilla extract or 2 inches vanilla bean

2 tablespoons cacao nibs

walnut or pecan pieces (optional)

directions

1. Peel and break the bananas into small pieces, roughly 4 pieces per banana.
2. Place them into a freezer-safe bag or container and freeze for at least 6 hours.
3. Allow bananas to thaw slightly. Place them into a food processor or high-speed blender with the vanilla and cacao ingredients and blend.
4. Top with walnut or pecan pieces (optional).
5. Serve immediately.

chocolate sauce

prep time: 5 minutes • **makes:** 1 cup

You never know when the perfect occasion for chocolate sauce will arise! This recipe requires just four ingredients, but creates a deep chocolate flavor. Use this sauce as a topping on ice cream or on our Black Bean Avocado Brownie (page 192) to make it even fudgier!

ingredients

1 cup cacao powder

½ cup coconut oil, melted

¼ cup maple syrup

pinch of salt

directions

1. In a small bowl, whisk together all ingredients until sauce is creamy.
2. To use as a drizzle sauce, stir in 1–2 teaspoons of warm water and stir until it reaches desired consistency.
3. Use the desired amount, and store in the fridge. Warm before each use.
4. To make hot chocolate, warm 6–8 ounces plant-based milk. Stir in 1–2 tablespoons chocolate sauce until melted.

* **Maple syrups come in four flavors:**
The two lighter versions, Golden and Amber, have a rich and delicate flavor. Dark and Very Dark have a deeper, more intense maple flavor and offer more beneficial antioxidants. Always buy pure maple syrup without added ingredients.

salted caramel sauce

prep time: 5 minutes • *cooking time:* 40 minutes • *makes:* 1 cup

Nothing beats the taste of a sweet and salty combination. Classic salted caramel sauce, however, contains butter and heavy cream. With just 5 ingredients, you can make an awesome dairy-free alternative that tastes just as sweet, buttery, and sticky as the sauce you remember. Drizzle it on ice cream and pancakes, and use it as dip for apples and pretzels!

ingredients

14 ounces coconut milk, full fat
¼ cup coconut sugar
½ teaspoon salt
1 teaspoon coconut oil
1 teaspoon vanilla extract or paste

directions

1. In a small saucepan, mix coconut milk, coconut sugar, and salt over medium high heat.
2. Bring to a boil, then lower the temperature to a light simmer.
3. Simmer for 30–40 minutes, stirring occasionally. As the sauce reaches the last 5 minutes of cooking, stir more frequently to incorporate the darker caramel bits from the bottom into the sauce.
4. Once the sauce has thickened enough to coat the back of a spoon, remove from the heat and stir in the coconut oil and vanilla extract.
5. Serve immediately or refrigerate.

resources

HOW-TO COOKING TIPS

Kombu

Chaey, C. (2016, October 17). *Are dried beans worth the effort?* Bon Appétit. www.bonappetit.com/story/dried-beans-worth-effort

Seidenberg, C. (2013, January 29). *Kombu, a nutritional powerhouse from the sea.* The Washington Post. www.washingtonpost.com/lifestyle/wellness/kombu-a-nutritional-powerhouse-from-the-sea/2013/01/29/aa4bb830-4ad4-11e2-a6a6-aabac85e8036_story.html

Soaking nuts

Saremi, L. (2020, October 1). *Why soak nuts?* Nature's Eats. www.natureseats.com/why-soak-nuts

Steaming

Overhiser, S., & Overhiser, A. (2020, July 2). *How to steam vegetables.* A Couple Cooks. Retrieved April 26, 2022, from www.acouplecooks.com/how-to-steam-vegetables

Steam cooking: A passion that matches the contemporary healthy diet pursuit. Eatwell101. (2012, February 5). www.eatwell101.com/cooking-with-steam-steam-cooking-food-steam-cooker-food-steamer

ABOUT THIS BOOK & MAMA'S FOOD PANTRY

Oats and oat flour

Palsdottir, H. (2022, April 4). *9 health benefits of eating oats and oatmeal.* Healthline. www.healthline.com/nutrition/9-benefits-oats-oatmeal

Almond flour

Raman, R. (2017, April 25). *Why almond flour is better than most other flours.* Healthline. www.healthline.com/nutrition/almond-flour

Einkorn flour

Covington, L. (2020, February 11). *What is einkorn?* The Spruce Eats. www.thespruceeats.com/what-is-einkorn-and-how-do-you-use-it-4705416

Gluten-free flour

Press, T. A. (2012, July 31). *Is your problem gluten? or faddish eating?* Gainesville Sun. www.gainesville.com/story/news/2012/07/31/is-your-problem-gluten-faddish/31603287007

Cassava flour

Link, R. (2021, November 23). *Cassava benefits and dangers.* Healthline. www.healthline.com/nutrition/cassava

Quinoa

Quinoa: The Inca's "Chisaya mama". Sharp Again Naturally. (2021, March 9). https://sharpagain.org/quinoa-the-incas-chisaya-mama

Beans

Warren, R. M. (2020, June 16). *Are beans good for you?* Consumer Reports. www.consumerreports.org/healthy-eating/are-beans-good-for-you-a8342413430

Chickpeas

Elliott, B. (2021, October 21). *10 health and nutrition benefits of chickpeas.* Healthline. www.healthline.com/nutrition/chickpeas-nutrition-benefits

Black beans

MediLexicon International. (n.d.). *Black beans: Health benefits, facts, and research.* Medical News Today. www.medicalnewstoday.com/articles/289934

Lentils

The Ultimate Guide to Lentils. Veecoco. (2021, January 28). www.veecoco.com/blog/ultimate-guide-to-lentils

Nuts and seeds

Carrie Dennett. (2016, March). *The wonders of nuts and seeds.* Today's Dietitian.
www.todaysdietitian.com/newarchives/0316p22.shtml

Almonds

Leech, J. (2018, September 6). *9 evidence-based health benefits of almonds.* Healthline. www.healthline.com/nutrition/9-proven-benefits-of-almonds

Walnuts

McCulloch, M. (2018, July 9). *13 proven health benefits of walnuts.* Healthline. www.healthline.com/nutrition/benefits-of-walnuts

Cashews

Cashews. Nuts.com. (n.d.). www.nuts.com/nuts/cashews

Flaxseeds

Amount of fiber in Flaxseed. Diet and Fitness. (n.d.). www.dietandfitnesstoday.com/fiber-in-flaxseed.php

Hemp seeds

Crichton-Stuart, C. (2018, September 11). *9 benefits of hemp seeds: Nutrition, health, and use.* Medical News Today. www.medicalnewstoday.com/articles/323037

Chia seeds

Chia. The Brasserie, Grand Cayman. (2017, June 27). www.brasseriecayman.com/superfood-series-chia

Pumpkin seeds

Brown, M. J. (2018, September 24). *Top 11 science-based health benefits of Pumpkin Seeds.* Healthline. www.healthline.com/nutrition/11-benefits-of-pumpkin-seeds

Sugar

Kubala, J. (2018, June 3). *11 reasons why too much sugar is bad for you.* Healthline. www.healthline.com/nutrition/too-much-sugar

Dates

Elliott, B. (2018, March 21). *8 proven health benefits of dates.* Healthline. Retrieved April 20, 2022, from www.healthline.com/nutrition/benefits-of-dates

Maple syrup

Canadian Maple Company. (n.d.). *Which grade of maple syrup is the healthiest?* www.puremaplesyrup.co/blogs/news/which-grade-of-maple-syrup-is-the-healthiest

Roberge, S. (2022, March 23). *Making the grade-the color and flavor of maple syrup.* University of New Hampshire Extension. www.extension.unh.edu/blog/2022/03/making-grade-color-flavor-maple-syrup

Coconut sugar

Gunnars, K. (2018, May 25). *Coconut sugar - a healthy sugar alternative or a big, fat lie?* Healthline. www.healthline.com/nutrition/coconut-sugar

Mathes, L. (2020, March 20). *5 reasons to love coconut palm sugar.* Healthy Goods. www.healthygoods.com/blog/5-reasons-to-love-coconut-palm-sugar

Ceylon cinnamon

Brennan, D. (2020, November 9). *Ceylon cinnamon: Health benefits, nutrients per serving, preparation information, and more.* WebMD. www.webmd.com/diet/health-benefits-ceylon-cinnamon

Raman, R. (2019, September 26). *6 side effects of too much cinnamon.* Healthline. www.healthline.com/nutrition/side-effects-of-cinnamon

Herbs and spices

What is the difference between organic and conventional herbs and spices. MCHEF®. (n.d.). www.mchef.com/blog/what-is-the-difference-between-organic-and-conventional-herbs-and-spices

Salt

BASc, J. S. (2020, March 5). *Top 5 health benefits of Celtic Sea Salt.* Organic Facts. www.organicfacts.net/celtic-sea-salt.html

Greenpeace International. (2019, May 25). *Over 90% of sampled salt brands globally found to contain microplastics.* Greenpeace International. www.greenpeace.org/international/press-release/18975/over-90-of-sampled-salt-brands-globally-found-to-contain-microplastics

Parker, L. (2021, May 3). *Microplastics found in 90 percent of table salt: Potential health impacts?* National Geographic. www.nationalgeographic.com/environment/article/microplastics-found-90-percent-table-salt-sea-salt

Peixoto, D., Pinheiro, C., Amorim, J., Oliva-Teles, L., Guilhermino, L., & Vieira, M. N. (2019, February 10). *Microplastic pollution in commercial salt for human consumption: A Review.* Estuarine, Coastal and Shelf Science. www.sciencedirect.com/science/article/pii/S0272771418300647

Oil, Coconut

MediLexicon International. (n.d.). *Combination of coconut oil and physical exercise could help reduce high blood pressure.* Medical News Today. www.medicalnewstoday.com/releases/289260#1

Oil, Olive

MediLexicon International. (n.d.). *Olive oil: Health benefits, nutritional information.* Medical News Today. www.medicalnewstoday.com/articles/266258

Weingus, L. (2015, August 11). *Yes, cooking with olive oil is perfectly safe.* HuffPost. www.huffpost.com/entry/olive-oil-explainer_n_55b925cce4b0a13f9d1b5143

RECIPES

Biggers, S. (2021, March 10). *Einkorn Wheat: An alternative for those that are most sensitive to most commercial wheat.* Backdoor Survival. www.backdoorsurvival.com/einkorn-wheat-an-alternative-for-those-that-are-sensitive-to-most-commercial-wheats

Hallal, F. (2021, September 10). *4 health benefits of pine nuts, according to science.* Healthline. Retrieved April 20, 2022, from www.healthline.com/nutrition/pine-nuts-benefits

Hallock, B. (2014, January 27). *To make a burger, first you need 660 gallons of water...* Los Angeles Times. www.latimes.com/food/dailydish/la-dd-gallons-of-water-to-make-a-burger-20140124-story.html

Hazra, A. (2018, October 17). The Millet Project. www.themilletproject.org/2015/02/26/the-millet-project

Hill, A. (2018, June 21). *Coconut aminos: Is it the perfect soy sauce substitute?* Healthline. Retrieved April 20, 2022, from www.healthline.com/nutrition/coconut-aminos

History of Pine Nuts & the people of the Great Basin. American Pine Nut Story | History of Pinon Pines | PineNut.com. (n.d.). www.pinenut.com/pinon-pinyon-history/value-nevada-forests.shtml

Julson, E. (2019, May 30). *6 benefits of liquid aminos (plus potential downsides).* Healthline. www.healthline.com/nutrition/liquid-aminos-benefits

Leech, J. (2021, March 18). *11 Scientifically Proven Health Benefits of Ginger.* Healthline. www.healthline.com/nutrition/11-proven-benefits-of-ginger

Lentils. The Nutrition Source. (2021, July 6). www.hsph.harvard.edu/nutritionsource/food-features/lentils

Link, R. (2021, November 23). *Cassava benefits and dangers.* Healthline. www.healthline.com/nutrition/cassava#nutrients

Link, R. (2017, June 30). *7 surprising health benefits of eggplants.* Healthline. www.healthline.com/nutrition/eggplant-benefits

MediLexicon International. (n.d.). *Cashews: Nutrition, health benefits, and Diet.* Medical News Today. www.medicalnewstoday.com/articles/309369#benefits

Millet flour. Millet flour nutrition facts and analysis. (n.d.). www.nutritionvalue.org/Millet_flour_nutritional_value.html

Millet notes - eden foods. (n.d.). www.edenfoods.com/articles/view.php?articles_id=122

Nutrition facts - rice flour. FreeFoodFacts. (n.d.). www.freefoodfacts.com/rice-flour

Petre, A. (2020, June 10). *Are cashews good for you? nutrition, benefits, and downsides.* Healthline. www.healthline.com/nutrition/are-cashews-good-for-you#nutrition

Pulses in your rotations - saskatchewan pulse growers. (n.d.). www.saskpulse.com/files/general/Pulses_in_rotations.pdf

Raman, R. (2019, May 24). *12 healthy ancient grains.* Healthline. www.healthline.com/nutrition/ancient-grains

Shockey, K. K., & Shockey, C. (2019, September 9). *Make your own miso.* Storey Publishing. www.storey.com/article/make-your-own-miso

Spritzler, F. (2020, January 27). *12 healthy foods that are high in iron.* Healthline. www.healthline.com/nutrition/healthy-iron-rich-foods

Warsh, J. (2018, August 31). *Modern Bread and Wheat Are Nothing like What Your Great Grandparents Ate.* Integrative Pediatrics and Medicine. www.integrativepediatricsandmedicine.com/modern-bread-wheat-gluten-is-nothing-like-the-bread-your-great-granparents-ate-the-science-and-history-behind-why-it-is-harming-our-health

What is cassava flour made from?: Bob's Red Mill. Bob's Red Mill Blog. (2021, December 28). www.bobsredmill.com/blog/healthy-living/what-is-cassava-flour-made-of

DR. GAIL MYERS

Myers, Gail. "Rhythms of the Land - Multimedia Film Project." *Dr. Gail Myers,* www.drgailmyers.com/projects

LEAH PENNIMAN

Figueroa, M., Penniman, L., Treakle, J., Pahnke, A., Calo, A., Iles, A., & Bowman, J. (2021, June 29). *Memo: Land access for beginning and disadvantaged farmers.* Data For Progress. www.dataforprogress.org/memos/land-access-for-beginning-disadvantaged-farmers

OCEAN & FISHERIES

Cho , R. |A. (2019, April 25). *Making fish farming more sustainable.* State of the Planet. Retrieved April 21, 2022, from https://news.climate.columbia.edu/2016/04/13/making-fish-farming-more-sustainable

Fisheries, N. O. A. A. (n.d.). *Understanding bycatch.* NOAA. www.fisheries.noaa.gov/insight/understanding-bycatch

Fish farming. Animal Welfare Institute. (n.d.). www.awionline.org/content/fish-farming

Getting your omega-3s vs. avoiding those pcbs.-the family HealthGuide. Harvard Health. (2004, April 1). www.health.harvard.edu/staying-healthy/getting-your-omega-3s-vs-avoiding-those-pcbsthe-family-healthguide

Goldfarb, B. (2016, June 1). *Struggling to rebuild the crashing Yukon River King Salmon run.* Anchorage Daily News. www.adn.com/rural-alaska/article/struggling-rebuild-crashing-yukon-river-king-salmon-run/2014/07/12

Lingel, G. (2020, December 12). *Fish farming: Harming oceans while poisoning people and the environment.* Sentient Media. www.sentientmedia.org/fish-farming

PCBS in farmed salmon. Environmental Working Group. (n.d.). Retrieved April 21, 2022, from www.ewg.org/research/pcbs-farmed-salmon#

Urry, A. (2015, July 31). *Everything you always wanted to know about fish farming but were afraid to ask.* Grist. www.grist.org/food/everything-you-always-wanted-to-know-about-fish-farming-but-were-afraid-to-ask

World Wildlife Fund. (n.d.). *What is overfishing? facts, effects and overfishing solutions.* WWF. www.worldwildlife.org/threats/overfishing

FACTORY FARMS

Aker, P. author B. A. (2021, July 28). *It's time for an urgent intervention in the Food System Ruining Our Climate.* Food & Water Watch. www.foodandwaterwatch.org/2021/07/21/its-time-for-an-urgent-intervention-in-the-food-system-ruining-our-climate

Animal Agriculture's impact on climate change. Climate Nexus. (2019, November 13). www.climatenexus.org/climate-issues/food/animal-agricultures-impact-on-climate-change

Factory farms make us sick. Food & Water Watch. (2021, April 22). Retrieved April 21, 2022, from www.foodandwaterwatch.org/2021/03/09/factory-farms-make-us-sick-2

Jones, A. (2022, March 30). *The cost of cheap chicken exposed in New York Times Video.* Species Unite. www.speciesunite.com/news-stories/the-cost-of-cheap-chicken-exposed-in-new-york-times-video

Patton, L. (2021, February 3). *The human victims of factory farming.* One Green Planet. www.onegreenplanet.org/environment/the-human-victims-of-factory-farming

Samuel, S. (2020, April 22). *The meat we eat is a pandemic risk, too.* Vox. www.vox.com/future-perfect/2020/4/22/21228158/coronavirus-pandemic-risk-factory-farming-meat

Shocking: New study finds public knows nothing about factory farming. Mercy For Animals. (2020, April 6). www.mercyforanimals.org/blog/shocking-new-study-finds-public-knows-nothing

GMOS & MONOCULTURES

Center for Food Safety and Applied Nutrition. (n.d.). *GMO crops, animal food, and beyond.* U.S. Food and Drug Administration. www.fda.gov/about-fda/fda-organization/center-food-safety-and-applied-nutrition-cfsan

Herbicide tolerant crops. Beyond Pesticides. (n.d.). www.beyondpesticides.org/resources/genetic-engineering/herbicide-tolerance

Hubbard, K. (2021, January 14). *The sobering details behind the latest seed Monopoly Chart.* Civil Eats. www.civileats.com/2019/01/11/the-sobering-details-behind-the-latest-seed-monopoly-chart

GMOs - top five concerns for family farmers. Farm Aid. (2019, May 2). www.farmaid.org/issues/gmos/gmos-top-5-concerns-for-family-farmers

Superweeds, secondary pests & lack of biodiversity are frequent GMO concerns. AgBioResearch. (2022, January 18). www.canr.msu.edu/news/superweeds-secondary-pests-lack-of-biodiversity-are-frequent-gmo-concerns

What is a GMO? - the non-GMO project. The Non-GMO Project - Everyone Deserves an Informed Choice. (2020, November 6). www.nongmoproject.org/gmo-facts/what-is-gmo

LAND ACCESS

Carlisle, L., de Wit, M. M., DeLonge, M. S., Calo, A., Getz, C., Ory, J., Munden-Dixon, K., Galt, R., Melone, B., Knox, R., Iles, A., Press, D., Kapuscinski, A. R., & Méndez, E. (2019, May 27). *Securing the future of US agriculture: The case for investing in new entry sustainable farmers.* University of California Press. https://online.ucpress.edu/elementa/article/doi/10.1525/elementa.356/112494/Securing-the-future-of-US-agriculture

Figueroa, M., Penniman, L., Treakle, J., Pahnke, A., Calo, A., Iles, A., & Bowman, J. (2021, June 29). *Memo: Land access for beginning and disadvantaged farmers.* Data For Progress. www.dataforprogress.org/memos/land-access-for-beginning-disadvantaged-farmers

Holloway, K. (2021, December 8). *How thousands of black farmers were forced off their land*. The Nation. www.thenation.com/article/society/black-farmers-pigford-debt

Newkirk II, V. R. (2020, June 16). *The Great Land Robbery*. The Atlantic. www.theatlantic.com/magazine/archive/2019/09/this-land-was-our-land/594742

Philpott, T. (2020, June 27). *White people own 98 percent of rural land. young black farmers want to reclaim their share.* Mother Jones. www.motherjones.com/food/2020/06/black-farmers-soul-fire-farm-reparations-african-legacy-agriculture

Rosenberg, N., & Stucki, B. W. (2020, December 17). *How USDA distorted data to conceal decades of discrimination against Black Farmers*. The Counter. www.thecounter.org/usda-black-farmers-discrimination-tom-vilsack-reparations-civil-rights

WHY ORGANIC?

Feldman, M., Ikerd, J., Watkins, S., Mitchell, C., Bowman, J., & Ostrander, C. R. (2021, June 19). *Regenerative farming and the green new deal*. Data For Progress. www.dataforprogress.org/memos/regenerative-agriculture-and-the-green-new-deal

Finck-Haynes, T. (2021, September 28). *"Insect apocalypse": Ace Hardware Acts on pesticides*. Friends of the Earth. Retrieved April 21, 2022, from www.foe.org/blog/insect-apocalypse-ace-pesticides

Lappé, A. (2022, March 21). *New study shows the growing risks of pesticide poisonings*. Civil Eats. Retrieved April 21, 2022, from www.civileats.com/2021/03/25/new-study-shows-the-growing-risks-of-pesticide-poisonings

Roseboro , K. (2018, May 18). *These farmers switched to organic after pesticides made their families sick*. Civil Eats. www.civileats.com/2018/05/11/these-farmers-switched-to-organic-after-pesticides-made-their-families-sick

The issue. Herbicide Free Campus. (n.d.). Retrieved April 21, 2022, from www.herbicidefreecampus.org/issue

United Nations. (2017, January 24). *United Nations report of the Special Rapporteur on the right to food* . Pesticide Action Network UK. www.pan-uk.org/site/wp-content/uploads/United-Nations-Report-of-the-Special-Rapporteur-on-the-right-to-food.pdf

THE WHYS OF A PLANT-BASED DIET

Scarborough, P., Appleby, P. N., Mizdrak, A., Briggs, A. D. M., Travis, R. C., Bradbury, K. E., & Key, T. J. (2014). *Dietary greenhouse gas emissions of meat-eaters, fish-eaters, vegetarians and vegans in the UK*. Climatic change. www.ncbi.nlm.nih.gov/pmc/articles/PMC4372775

index

about the authors

MACKENZIE FELDMAN

Mackenzie Feldman is an environmental activist from Honolulu, Hawai'i. Growing up on Oahu, Mackenzie saw the effects that corporate agribusiness and the resulting pesticide exposure have had and continue to have on her community. Simultaneously, Mackenzie witnessed the power of the Hawaiian food sovereignty movement. She graduated from the University of California, Berkeley in Spring 2018 with a Bachelor of Science degree in Society and Environment and a minor in Food Systems. She is the Founder and Executive Director of Herbicide-Free Campus, an organization that works with students and groundskeepers around the country to advocate for an end to the spraying of synthetic herbicides at schools and a transition to organic land management. Her campaign resulted in the entire University of California system going glyphosate-free, and Mackenzie worked with a coalition to get herbicides banned from every public school in the state of Hawai'i. Mackenzie is also a Food Research Fellow for Data For Progress, a Rachel Carson Council Fellow and received the Brower Youth Award for her work with Herbicide-Free Campus.

KATHY FELDMAN

Kathy Feldman's quest for plant-based eating began early on in life. As an animal lover, she wanted to stop eating meat in the 4th grade, but had to wait until the 11th grade when she could cook for herself to begin the journey of eating vegetarian. As a teen, she got a hold of *Diet for a Small Planet* by Frances Moore Lappé, which made her more aware of the environmental impact meat consumption had on the planet. It wasn't until Kathy read the book *Diet for a New America* by John Robbins, did she learn about the horrific treatment of animals in factory farms, which validated her decision to not eat meat. When genetically modified papayas reached the market in Hawai'i, Kathy set out to research GMOs. She was shocked to learn that corn and soy seeds are engineered to withstand the spraying of herbicides. In 2016, Kathy and Mackenzie set out to learn more about the food system, visiting 10 farms in Cuba and 20 organic farms in California and Oregon, keeping a blog about their experiences along the way. Since then, the two decided to write this book to share plant-based recipes and highlight some of the inspiring farms and food system leaders they met. Kathy is a candidate for State House, District 19 on Oahu, Hawai'i, challenging an incumbent who is the Chair of the Agricultural Committee and a big supporter of industrial agriculture. Kathy is running to stand up for what is right and champion environmental protection, affordable housing, and better education.

how to get involved
Check out these resources below!

BOOKS

- Animal, Vegetable, Junk: A History of Food, from Sustainable to Suicidal
- Behind the Kitchen Door
- Bite Back: People Taking On Corporate Food and Winning
- Eating NAFTA: Trade, Food Policies, and the Destruction of Mexico
- Eating Tomorrow: Agribusiness, Family Farmers, and the Battle For the Future of Food
- Farming While Black: Soul Fire Farm's Practical Guide to Liberation on the Land
- Freedom Farmers: Agricultural Resistance and the Black Freedom Movement
- Fresh Fruit Broken Bodies: Migrant Farmworkers in the United States
- Healing Grounds: Climate, Justice, and the Deep Roots of Regenerative Farming
- Salt Sugar Fat: How the Food Giants Hooked Us
- The Farm Bill: A Citizen's Guide

Of course there are many more resources out there, but here are some of our favorites!

ORGANIZATIONS

- Beyond Pesticides
- Black Urban Growers Association
- Center for Food Safety
- Family Farm Defenders
- Federation of Southern Cooperatives
- Food and Water Watch
- Food Chain Workers Alliance
- Friends of the Earth
- HEAL Food Alliance
- The Land Loss Prevention Project
- National Black Food and Justice Alliance
- National Sustainable Agriculture Coalition
- National Young Farmers Coalition
- Pesticide Action Network
- Real Food Media
- The Northeast Farmers of Color Land Trust
- Soul Fire Farm

MEDIA OUTLETS

- Civil Eats
- Food and Environment Reporting Network
- Food Tank
- Modern Farmer
- Mother Jones
- Whetstone Media

www.ingramcontent.com/pod-product-compliance
Lightning Source LLC
Chambersburg PA
CBHW040439150626
46551CB00025B/137